# The Handbook of Environmental Chemistry

Volume 1    Part E

Edited by O. Hutzinger

# The Natural Environment and the Biogeochemical Cycles

With contributions by
G. H. Dury, R. Eiden, J. R. Holton, L. Johnson

With 66 Figures

Springer-Verlag Berlin Heidelberg GmbH

Professor Dr. Otto Hutzinger

University of Bayreuth
Chair of Ecological Chemistry and Geochemistry
Postfach 101251, D-8580 Bayreuth
Federal Republic of Germany

ISBN 978-3-662-15071-9     ISBN 978-3-540-39463-1 (eBook)
DOI 10.1007/978-3-540-39463-1

© Springer-Verlag Berlin Heidelberg 1990
Originally published by Springer-Verlag Berlin Heidelberg New York in 1990.
Softcover reprint of the hardcover 1st edition 1990

Typesetting: Macmillan India Ltd., Bangalore-25
Offsetprinting: Color-Druck Dorfi GmbH, Berlin;
2152/3020-543210

# Preface

Environmental Chemistry is a relatively young science. Interest in this subject, however, is growing very rapidly and, although no agreement has been reached as yet about the exact content and limits of this interdisciplinary subject, there appears to be increasing interest in seeing environmental topics which are based on chemistry embodied in this subject. One of the first objectives of Environmental Chemistry must be the study of the environment and of natural chemical processes which occur in the environment. A major purpose of this series on Environmental Chemistry, therefore, is to present a reasonably uniform view of various aspects of the chemistry of the environment and chemical reactions occurring in the environment.

The industrial activities of man have given a new dimension to Environmental Chemistry. We have now synthesized and described over five million chemical compounds and chemical industry produces about one hundred and fifty million tons of synthetic chemicals annually. We ship billions of tons of oil per year and through mining operations and other geophysical modifications, large quantities of inorganic and organic materials are released from their natural deposits. Cities and metropolitan areas of up to 15 million inhabitants produce large quantities of waste in relatively small and confined areas. Much of the chemical products and waste products of modern society are released into the environment either during production, storage, transport, use or ultimate disposal. These released materials participate in natural cycles and reactions and frequently lead to interference and disturbance of natural systems.

Environmental Chemistry is concerned with *reactions in the environment*. It is about distribution and equilibria between environmental compartments. It is about reactions, pathways, thermodynamics and kinetics. An important purpose of this Handbook is to aid understanding of the basic distribution and chemical reaction processes which occur in the environment.

Laws regulating toxic substances in various countries are designed to assess and control risk of chemicals to man and his environment. Science can contribute in two areas to this assessment: firstly in the area of toxicology and secondly in the area of chemical expsoure. The available concentration ("environmental exposure concentration") depends on the fate of chemical compounds in the environment and thus their distribution and reaction behaviour in the environment. One very important contribution of Environmental Chemistry to the above mentioned toxic substances laws is to develop laboratory test methods, or mathematical correlations and models that predict the environmental fate of new chemical compounds. The third purpose of this Handbook is to help in the basic understanding and development of such test methods and models.

The last explicit purpose of the handbook is to present, in a concise form, the most important properties relating to environmental chemistry and hazard assessment for the most important series of chemical compounds.

At the moment three volumes of the Handbook are planned. Volume 1 deals with the natural environment and the biogeochemical cycles therein, including some background information such as energetics and ecology. Volume 2 is concerned with reactions and processes in the environment and deals with physical factors such as transport and adsorption, and chemical, photochemical and biochemical reactions in the environment, as well as some aspects of pharmacokinetics and metabolism within organisms. Volume 3 deals with anthropogenic compounds, their chemical backgrounds, production methods and information about their use, their environmental behaviour, analytical methodology and some important aspects of their toxic effects. The material for volumes 1, 2, and 3 was more than could easily be fitted into a single volume, and for this reason, as well as for the purpose of rapid publication of available manuscripts, all three volumes are published as a volume series (e.g. Vol. 1; A, B, C). Publisher and editor hope to keep the material of the volumes 1 to 3 up to date and to extend coverage in the subject areas by publishing further parts in the future. Readers are encouraged to offer suggestions and advice as to future editions of "The Handbook of Experimental Chemistry".

Most chapters in the Handbook are written to a fairly advanced level and should be of interest to the graduate student and practising scientist. I also hope that the subject matter treated will be of interest to people outside chemistry and to scientists in industry as well as government and regulatory bodies. It would be very satisfying for me to see the books used as a basis for developing graduate courses on Environmental Chemistry.

Due to the breadth of the subject matter, it was not easy to edit this Handbook. Specialists had to be found in quite different areas of science who were willing to contribute a chapter within the prescribed schedule. It is with great satisfaction that I thank all authors for their understanding and for devoting their time to this effort. Special thanks are due to the Springer publishing house and finally I would like to thank my family, students and colleagues for being so patient with me during several critical phases of preparation for the Handbook, and also to some colleagues and the secretaries for their technical help.

I consider it a privilege to see my chosen subject grow. My interest in Environmental Chemistry dates back to my early college days in Vienna. I received significant impulses during my postdoctoral period at the University of California and my interest slowly developed during my time with the National Research Council of Canada, before I was able to devote my full time to Environmental Chemistry in Amsterdam. I hope this Handbook will help deepen the interest of other scientists in this subject.

Otto Hutzinger

This preface was written in 1980. Since then publisher and editor have agreed to expand the Handbook by two new open-ended volume series: Air Pollution and Water Pollution. These broad topics could not be fitted easily into the headings of the first three volumes.

All five volume series will be integrated through the choice of topics covered and by a system of cross referencing.

The outline of the Handbook is thus as follows:
1. The Natural Environment and the Biogeochemical Cycles
2. Reactions and Processes
3. Anthropogenic Compounds
4. Air Pollution
5. Water Pollution

Bayreuth, July 1989                                    Otto Hutzinger

# Contents

# List of Contributors

Professor G. H. Dury
46 Woodland Close
Risby
Bury St. Edmunds
Suffolk IP28 6QN
England

Prof. Dr. Reiner Eiden
Universität Bayreuth
Institut für Geowissenschaften
Abteilung Meteorologie
Postfach 30 08
Universitätsstr. 30
8580 Bayreuth, FRG

Prof. Dr. James R. Holton
University of Washington
Department of Atmospheric
Science, AK-40
Seattle, Wahington 98195
USA

Dr. Lionel Johnson
Government of Canada
Fisheries and Oceans
Freshwater Institute
501 University Crescent
Winnipeg, Manitoba R3T 2N6
Canada

# The Thermodynamics of Ecosystems

*Lionel Johnson*

Freshwater Institute, Department of Fisheries and Oceans, Western and Arctic Region, 501 University Crescent, Winnipeg, R3T 2N6, Canada.

## Summary

The properties of autonomous Arctic lake ecosystems, within the context of the whole living World, provide the basis for the development of a thermodynamic theory of the biosystem. The biosystem is now undergoing a process of internal reordering. Energy flow is being accelerated at all levels in virtually all ecosystems in order to support increased biomass at the terminal level. The thermodynamic constraints, particularly the need to maintain near-symmetry, impose limits on the capacity of the system to respond. This has important consequences for environmental management.

## Introduction

'The basic problem is how to think about ecosystems and how to place them within the scheme of known systems.'

*(Engleberg and Boyarsky 1979)*[1]

Science may be defined as the attempt to make generalized statements about the universe in order to simplify concepts and understand the workings of nature. However, apart from the general statement on evolution by natural selection as forged by Charles Darwin [2] and subsequently hammered into the "Modern Synthesis" by the Neo-Darwinists, few generalizations on biology have been possible and none has been fully integrated into the general body of scientific knowledge. The living world has existed as a cryptogram in which only a few words have been deciphered, leaving the fundamental message unresolved. Yet without the fundamental theory that decipherment would bring, our attempts to rectify the apparently disastrous course of events in a rapidly changing environment are likely to be of little avail. "Ecology must have a strong theoretical base before it can advance significantly" [3] or, in the often quoted but still vivid phrase of Watt [4];

"If we do not develop a strong theoretical core that will bring all parts of ecology back together we shall all be washed out to sea on an immense tide of unrelated information."

Without such a scientific synthesis bringing the living world into harmony with our wider understanding of physics and chemistry it is difficult to develop principles concerning the effect of environmental change on living organisms. If the doubt and uncertainty remain we shall continue to try and control, rather than work in concert with nature.

In recent years the intensive struggle to decipher the cryptogram has had relatively little success. Attacks on the problem have ranged from ecological studies and experiments on the effects of competition to the development of mathematical models, but with little sign or progress in either direction.

"Ecology," McIntosh [5] states, "is now undergoing a period of retrenchment. This follows on nearly three decades of heady belief on the part of some ecologists, newly ventured into the maze of community ecology, that communities are structured in an orderly and predictable manner, and of others that information theory, systems analysis, and mathematical models would transform ecology into a "hard" science."

Unfortunately, there has been considerable division within ecology. Apart from the wide split between the theoretical/mathematical approach and that of the more practical field ecologist, there has also been a division within field-oriented ecologists themselves. McIntosh [6–8] attributes these differences to the development of "Invisible colleges" (an epithet first applied by Robert Boyle to the British Royal Society at the time of its formation in the seventeenth century), frequently established around one eminent authority. The difficulty experienced in bridging the schisms between the colleges, McIntosh [6] believes, is due to strong philosophical biases.

Disillusionment, consequent upon lack of real progress in understanding ecosystem mechanics, caused Hirata and Ulanowicz [9] to state:

"The time for regarding ecosystems as collections of deterministic processes (as has been the convention in most simulation modelling) appears to be over. There is every reason to believe that individual processes actually change in nature as a result of their mutual interaction with other processes in the same community. Such circumstances would require that ecosystems be viewed as a unit with identifiable properties which exhibit some macroscopic regularities. (Dare we say laws?)."

Not only dare we say laws, I believe that we must say laws if, as biologists, we wish to remain within the scientific community. These laws, like the laws of statistical mechanics, may only be observable at the uppermost level of generality, while events at the local level appear to be largely without pattern. This presupposes a world view which has obvious difficulties both for observation and experiment. Nevertheless, we must proceed in the faith that the cryptogram contains the laws embedded in the message. The difficulties of decipherment, I believe, are a consequence of the nature of the laws not an indication of their nonexistence.

The differences that have arisen in ecology arise mainly from differences in opinion as to the existence and nature or organizing forces. Specifically it is questioned whether organizing forces exist above and beyond those inherent in the genetic make-up of the individual species, and regularities of the environment. Amplification of these differences has led to a dualism, a separation of the organism from its environment and the treatment of the two factors as separate entities. This dualism has tended to supplant the original, pre-Darwinian "unified organism-environment whole (synergism)" [10]. Darwin, although himself subscribing to the holistic approach, emphasised the struggle of the evolving organism to survive. This led his followers to isolate and emphasize the importance of the individual organism as expressed in "survival of the fittest."

In reaction to this dualism between organism and environment, there is now a strong movement reverting to a more holistic approach in which biological entities above the level of the individual are regarded as part of a system in the full sense of the word. The appropriate approach to the investigation of such an entity, according to Patten [10–11], Patten and Odum [12] and Patten and Auble [13] is that of 'systems analysis.' This is, in essence, a more formal perspective on the older concepts advocated by Forbes [14], Clements [15–16], Tansley [17], Odum [18–19] and Dunbar [20]. The central notion is that properties of whole systems must be examined and analysed if the internal workings are to be understood.

A system may be defined as a complex whole, a set of connected things or parts, or an organized body of material or immaterial things. A system is thus isolated from other systems. Connected to other systems it becomes a subsystem. Within a system a change in one component results in changes in all other components.

With this definition of a system in mind, Tansley [17] coined the term *ecosystem* to avoid what he felt were strong objections to the term *community*. A community, he maintained, implies a mutual goal scarcely apparent in the workings of nature of the local level. An ecosystem, Tansley defined as

"   . . . the whole system (in the sense of physics), including not only the organism-complex, but the whole complex of physical factors forming what we call the environment of the biome–the habitat factors in the widest sense."

In the physical world, two types of basic system are distinguished: an *isolated* system in which entropy increases with time, reaching a maximum at thermodynamic equilibrium, and an *open* system in which energy and materials are exchanged with the external world. A special case of the open system of great importance in biology is the *closed* system: one in which only energy and entropy are exchanged with the external universe [21]. (The internal inconsistency of these terms is unfortunate and has caused much confusion in biological writings). An isolated system eventually reaches thermodynamic equilibrium and is of little further interest biologically. On the other hand, the world biological system, within the biosphere boundaries, functions as a closed system, maintained by a free energy gradient. Free energy is energy capable of doing work. In that the biosphere boundaries are permeable to energy, but not biological material, such a system is incapable of reaching thermodynamic equilibrium. It therefore functions as a *non-equilibrium* system. Free energy in the form of short-wave radiation from the sun is temporarily stored in material cycles, then ultimately dissipated to the universe as long-wave radiation when it can no longer perform useful work.

Within the world system an ecosystem, functioning locally, is not completely isolated from its neighbours; it is therefore a sub-system of the greater whole. These subsystems develop within partial boundaries of a physical or biological nature. The boundaries endow every ecosystem with a varying degree of autonomy. As Tansley [17] states: "The more relatively separate and autonomous the system, the more highly integrated it is, and the greater the stability of its dynamic equilibrium." Every identifiable ecosystem is thus the focus of a concentric series of partial boundaries which gradually expand outwards until the limits of the biosphere are reached.

The general nature of the organizing forces within the ecosystem, and at the world level, must be capable of formulation in thermodynamic terms if it is to fall within the general purview of science [22–27]. Fundamentally, the subject matter of biology is therefore organization. Organization, according to Pittendrigh [28] (p.391), demands an 'end- point' or 'goal', toward which a system proceeds. This, at first sight, implies a teleological interpretation of events, anathema to many biologists, but as Mayr [28] points out, many physical effects, such as gravity, lead to an end-point and are therefore 'teleomatic,' others, such as the workings of a computer programme, or the development of an embryo into an adult organism, are 'teleonomic.' At the system level, organization implies an end-point, or specific configuration, to which a system proceeds. This end-point can be reached only when environmental conditions follow a relatively constant cycle and the species from which the ultimate complement is derived remain the same. The end-point is thus a recognizable stationary state. However, there is little general agreement on whether a stationary or stable state does exist, and certainly in the present disturbed state of the biosystem such a stable state is likely to be most difficult to find. Thus the validity of terms such as "equilibrium," "dynamic equilibrium," or "climax," has been called into question.

The organizing forces, according to Tansley are:

"on the one hand the total net action of the effective environmental factors, on the other the combined actions of the individuals themselves," or, as recently stated by Oksanen [29]:

". . . the struggle for existence is . . [the ecosystem's] . . sole organizing principle , and the physical environment is its only external constraint."

However, these statements as they stand are not reducible to laws compatible with those of physics and chemistry because the nature of 'the struggle for existence' is not defined in the appropriate terms.

In our highly disturbed modern world, very few systems have the opportunity to approach a stationary state, should such be the goal or end-point of change. Even in times before man's massive intervention, climatic or other environmental change tended to alter continuously the background against which systems develop. Nevertheless, over geological time, there were intervening periods in which relative stasis was attained [30–35]. In today's biosystem, apparently stable autonomous ecosystems can still be found in the ecological hinterlands [36].

Before ecosystem operation can be interpreted in thermodynamic terms it seems essential to establish an idealized reference system, comparable with the isolated system of thermodynamics. Such a reference system will be an idealized stationary state. The biological reference system is to be sought in an autonomous ecosystem that has been free from disturbance within its recent history. From its characteristics it should then be possible to identify patterns indicating the nature of the organizing forces. In the intervening period it is necessary to hold in abeyance all the extraneous concepts and assumptions that clutter our view of ecosystem mechanisms. It is therefore desirable to start afresh with the observed evidence from real ecosystems, and attempt to disentangle processes on the basis of known and accepted physical postulates. We must put aside for the moment all attempts to convert biology to a "hard" science through application of Newtonian principles as well as current notions of "the struggle for existence," "fitness," "survival," "competition," "group-selection," "individual-selection," and "kin-selection."

The term *equilibrium* has been used in so many contexts that it is preferable to confine its usage to *thermodynamic equilibrium*—a state of maximum entropy. Then by definition ecosystems are *non-equilibrium systems* existing *far-from-equilibrium*. However, this does not preclude ecosystems from reaching a stationary state, stable within the environmental conditions they have experienced over their recent past history.

The term *biosystem* is used to refer to the whole world system in preference to the more commonly used *biosphere*; the biosphere is conceived as the space occupied by the biosystem. This usage emphasises the system nature of the living world and the bounded nature of the space within the biosphere.

## Emergent Properties of the Biosystem

From fifth century B.C., when Herodotus, for the first time recorded, discussed the concept that ultimately became the 'Balance of Nature' [37], biologists have struggled to understand the principles of biological organization. Few additional generalizations were made until the perspicacious Captain James Clark Ross, R.N., during his voyage to the Antarctic in 1839–43, after three previous voyages to the Arctic, recognized a general similarity between the Arctic and Antarctic faunas associated with the sea ice [38]. Although different species were involved, the basic structure of the marine/ice ecosystem of Arctic and Antarctic was comparable, with invertebrates feeding on the under-ice surface being preyed upon by small fishes, which in turn supported the marine mammal and bird fauna living above the ice but feeding below it. Different aggregations of organisms evidently respond similarly to similar environments in a manner indicating underlying organizational processes.

Soon after Captain Ross' observation, which perhaps did not receive the acclaim it deserved, came the most important generalization to date, that of Charles Darwin's theory of evolution through natural selection. This generalization again emerged from observations made during the world voyage of H.M.S. "Beagle". About the same time, Alfred Russell Wallace [39], likewise as a result of voyages of collection and exploration in the tropics, came to conclusions comparable with those of Darwin on evolution and natural selection.

A second major generalization of Wallace [40] arose from his work on plant and animal geography. He concluded that species richness (the number of species in a given area) increases from high latitudes toward equatorial regions. The reasons for this gradient have remained something of an enigma, and have given rise to much discussion and speculation [41–47].

As ecology emerged as a separate discipline within biology, about the turn of the century, [8], new ideas were introduced. These were concepts of community structure [48] and succession; succession being the more–or–less orderly development of vegetation on a new site [49, 50].

Following Ross' observations, similar correspondences were identified in many parts of the world, with very different organisms filling comparable ecological roles. These correspondences occur in both terrestrial and aquatic systems [51–52]. Broad-leafed forests develop in moist, warm conditions, but in harsher conditions coniferous forests are more prevalent. Grasslands, invariably grazed by dominant herbivores, occur in drier, colder or more variable climates. The dominant herbivore, may be a bird, such as the moa, formerly existing in New Zealand, reptile, as in the giant tortoise of Aldabra, or one, or many of the ungulate species that occupy, or formerly occupied the grasslands of Europe, Africa and the Americas. Fish faunas of temperate rivers follow a consistent pattern from headwaters to sea-level [53]. Tropical freshwaters are extremely rich in species; nevertheless, the structure of the ecosystem tends to follow certain general principles in which several species appear to be ecologically equivalent [54, 55].

In their experimental work on the recolonization of artificially defaunated small islands off the coast of Florida, Simberloff and Wilson [56, 57] found that the insect fauna that re-established itself was not identical with the original one, but that the overall structure was similar, with different species filling equivalent ecological roles.

Such tantalizing glimpses of evident organisation at the world level are sufficient only to support an initial faith in the existence of organizing forces. Unfortunately, much of the text of the cryptogram (of which we are an integral part), particularly those parts found in the geological record and in current ecosystem structure, was destroyed before it could be described and recorded. There is also a problem of scale; it evident from the macroscopic pattern of the symbols, that some message is being transmitted, but the regularity observed at the macroscopic level almost invariably disappears when the text is examined at the microscopic level. In the same way it would be impossible to discern the gas laws from a description of the path of a single gas molecule. Only through repeated switching from the macroscopic perspective to the microscopic, from holistic to reductionist point of view, is it possible to arrive at an answer that conforms to both perspectives.

## Arctic Lake Ecosystems

Initial stimulus for the development of a general hypothesis was provided by long-term investigations on arctic and sub-arctic lakes in the Canadian Northwest Territories [58–60]. Such lakes form ideal laboratories for the investigation of whole systems. They have been undisturbed since their formation in the late Pleistocene, therefore, they have had sufficient time to assume a stationary state should such be their end-point. For 9 to 10 months each year they exist as completely closed ecosystems, sequestered from the external world by a layer of ice up to 2.5 m thick. For the remaining 2 to 3 months exchange of material with the external world is quite limited. The food-chains are simple and relatively linear. From a thermodynamic viewpoint they closely represent the world in microcosm. In fact, they fulfil the precept of Platt [61] that the investigation of complex phenomena should begin with the simplest system available that contains all the variables under consideration.

Within the Northwest Territories, lake ecosystems exist in their simplest form in the most northerly regions. Lake Hazen (Lat. 81° 50′N) for example, at the north end of Ellesmere Island, is the world's largest lake entirely north of the Arctic Circle. It has a sparse benthic fauna of aquatic mites (Order Acarina) and midge larvae (Family Chironomidae) and a planktonic invertebrate fauna of one species of copepod, *Cyclops scutifer*, and two species of rotifer [62]. The terminal species in this simple food-chain is the single fish species, Arctic charr, *Salvelinus alpinus* [59–60]. In lakes of islands to the south, the number of fish and invertebrate species increases, but within one island this is not a strict geographic relationship: lake morphology is evidently of great importance in determining the species complement. In one area of Victoria Island, for example, adjacent lakes may contain different species sub-sets, the actual species complement in a specific lake being unrelated to latitude, elevation or drainage pattern. The most economical explanation is that all lakes in the region under consideration had the same set of species available to them when the land emerged from the sea in the post-Pleistocene [63], the present species composition then being determined by inter-specific interaction. At the present time, the complete species set is found only in large lakes near sea level [59]. As lakes of differing morphologies became isolated, only certain species subsets were able to survive. This is evidently a case of competitive elimination, the species surviving being dependent on the specific environmental conditions. This results in a distributional mosaic in which one lake may contain a single fish species (invariably Arctic charr) in close proximity to another in which lake trout, *Salvelinus namaycush*, lake whitefish, *Coregonus clupeaformis*, and ciscos, *Corgeonus spp.*, are present. An additional variant is a lake containing Arctic charr in addition to lake trout and coregonids. Further south, on the mainland, Arctic charr is gradually eliminated, the dominant species being lake trout and lake whitefish, with an increasing number of species of less relative importance [57–58].

The population structure of the dominant or co-dominant fish species in every lake, irrespective of whether there is a single or multiple species fish community, exhibits similar characteristics. The fish are of considerable age (Arctic charr up to 25–30 y, and lake trout up to 60 y), large size (frequently up to 100 cm), of uniform size (despite wide variation in age), and of indeterminate age at death. There is also a remarkable dearth of juveniles. Not only is this structure universal in the lakes examined, it has been shown to exist indefinitely, or at least over the period of observation of a single lake, which now exceeds 25 years. Because the populations maintain a constant configuration over long time periods, they must be considered stable in the face of environmental fluctuations to which they are normally exposed.

Significantly, uniformity in size is, to a considerable extent, independent of age. That is, although fish only increase in size with time (resulting in an overall correlation of size and age), the individual growth pattern is subject to a degree of modification which ensures that the fish in a particular size class exhibit a wide range of age. In a single species fish community, uniformity in size (with great variation in age), and the apparent dearth of immatures, implies intraspecific population regulation. Food habits show that cannibalism is not responsible for this configuration, although in some cases it may be a contributary cause [64].

The pattern of age- and length-frequency distribution of the northern fish populations examined were all essentially similar, although the actual modal size and modal age varied considerably from lake to lake [58, 60]. This is interpreted as an adaptation to the local environment, or 'fine-tuning' to the prevailing conditions.

The general conclusion was reached that the geographic distributional pattern indicates that competition (competitive elimination) is important in determining the species complement, but the actual structuring of individual populations is the result of mechanisms internal to the population.

Experimental manipulation of single and multiple species fish populations showed that, following disturbance of an external nature (i.e. a disturbance that does not change the driving variables [65], these populations, given sufficient time, return to the ground state in a well-damped manner [60, 66]. Return to the initial ground state implies the existence of an end-point or goal, thereby demonstrating the existence of an organizing force. The absence of oscillation during recovery indicates that the species population, or ecodeme, functions as a coherent, organized entity, not as a group of individuals. The well-damped recovery indicates the existence of an interactive mechanism effecting population control at all stages of the process. Therefore these mechanisms *are not density dependent*.

The results express two apparent anomalies with respect to generally accepted theory. First, the general configuration of the fish stocks observed is contrary to that adopted in much fish population dynamics theory [67, 68]. In these works it is assumed that individual length is correlated with age (which, in general terms is indisputable), therefore, as numbers decline with age, the length-frequency assumes a negative exponential distribution. However, the fish populations in these undisturbed northern lakes exhibit a bell-shaped size-frequency distribution, approximating that of a normal, or Gaussian curve. This distribution pattern is comparable to that of a bird or mammal population, where size is generally considered to be under genetic control.

Second, and particularly challenging at the start of these investigations when the "diversity-stability" hypothesis [69, 70] was at the height of its popularity [71], was that these simple northern fish population were evidently stable. The diversity-stability hypothesis, based on analogy with information theory, stated that the greater the diversity of an ecosystem, the greater the number of interconnections, and the greater the stability of the whole. Diversity in this context is a measure of species richness and equitability (the evenness with which species are distributed). MacArthur [69] postulated that highly diverse systems such as tropical rain forests are relatively stable but

"Where there is a small number of species (e.g. in arctic regions) the stability condition is hard or impossible to achieve.'

The conclusion that simple systems fluctuate was based on firm evidence of fluctuation in small mammal populations of the tundra and northern forest edge [72, 73, 74]. This was a direct conflict of evidence in that fluctuating populations occurred on lands surrounding lakes in which stable populations existed. In due course the diversity-stability hypothesis was severely criticized on thermodynamic grounds by Goodman [75] and interest in it lapsed. This, however, does not negate the evidence on which the hypothesis was based.

To resolve the conflict between lake and tundra populations, it was concluded that an autonomous ecosystem, developing within relatively impermeable boundaries, assumes a stationary state irrespective of its diversity or simplicity. The arctic lake ecosystem is highly autonomous, therefore it assumes a stable state, whereas, conversely, the tundra system is of immense extent, internally quite variable and wide open to the annual migration of birds and mammals. In addition a lake is highly buffered from environmental fluctuation by the physical properties of water, whereas the terrestrial ecosystem is subjected to severe climatic fluctuations both within and between years.

It is therefore postulated that all autonomous ecosystems, irrespective of their diversity, tend to assume a stable state, within the boundary constraints of the system, the regularity of the energy input and the genetic make-up of their species populations.

### Generality of the Findings

The important question to be answered is: Does the structure of arctic fish populations represent a proposition of general validity, or is it specific to arctic populations? A survey of the literature indicated that such a state was indeed general (Johnson [36, 76],). The main points may be summarized:

## i) The Giant Land Tortoise of Aldabra Atoll

Aldabra Atoll, in the Indian Ocean, is a dependency of the Republic of the Seychelles. Geographically, the various islands of the atoll are the inverse of an arctic lake: they are tropical, terrestrial, surrounded by water not land, and the dominant species is a reptile, the giant land tortoise, *Geochelone gigantea*. The distance from other land ensures a high degree of autonomy. The tortoise populations have been intensively studied by a number of workers: Stoddart and Wright [77], Gaymer [78], Grubb [79]; Bourn and Coe [80], Swingland and Coe [81], Swingland and Lessels [82], and Gibson and Hamilton [83]. Stoddart and Wright [77] state that ". . . Aldabra is one of the least disturbed of all low-latitude islands, and for historical and environmental reasons possesses and exceptionally rich and interesting fauna and flora."

The tortoise populations were decimated in the late nineteenth century to supply the needs of the 'tortoise-shell' industry, but subsequently have remained undisturbed. The islands also suffered the introduction of a number of exotic species such as cats, goats and dogs, but for the past 70 years they have been maintained as a nature reserve. The only industrial activity is the cultivation of coconuts on a limited scale. At the present time land tortoises are the only large land animals, apart from a few goats on South Island [78].

Length-frequency distributions of the various tortoise populations on individual islands, or parts of larger islands, are invariably unimodal. The modal value of the length along the carapace is usually about 750 mm. Males may reach 1000 mm; females are generally under 800 mm. All the samples collected in a systematic manner showed an apparent under-representation of small tortoises. Grubb [79] states "Because of the paucity of small tortoises in these samples a special effort was made to measure animals of about 20 cm carapace length or less." Bourn and Coe [80] comment "The shape of the 'pyramid' [of the population size structure] is indicative of a declining population, in which the pre-reproductive size classes form a relatively small proportion."

Giant land tortoises are known to live to a great age, estimated by Gibson and Hamilton [83] to be up to 60 years. However, ages, as determined from growth rings on the scutes, are difficult to determine after 20–25 years. On South Island, the modal size of 750 mm included tortoises between 13 and 24 years old. Similarly for Middle Island, Grubb [79] concluded that the modal size of 700–750 mm consisted of animals from 15 to 26 years. On Anse Mais, Gaymer [78] could find no tortoises under five years old and concluded that the population was not breeding. The population structure of these tortoises is remarkable similar to that of lake trout in Great Bear Lake, in both size and age.

## ii) The Virgin Forests of Europe

A detailed review of the structure of the virgin forests of Europe was completed by Jones [84]. A negative exponential diameter-distribution is considered by some foresters to be the ideal configuration for maximum production. This is referred to as 'selection' forest. Jones continues:

> "A far commoner form of uneven-aged forest [than the 'selection' type forest] has very much the appearance of an even-aged high forest. It is perhaps too much to say that all ages are present, but although old stems predominate, the dominant stand includes a wide range of ages. In an example of Bosnian beech forest given by Cermak [1910], quoted in Ref [84], the age range was 200 years, though nearly 70% of the stems were between the limits of 170 and 200 years. The youngest stems were 90–100 years old. Regeneration in such forests often appears to be absent, and there is a striking deficiency of smaller stems, giving a diameter distribution which may approach an even-aged stand. This, however, does not necessarily mean that the forest is not reproducing itself and maintaining its present structure: the condition doubtless arises through the relatively small fraction of the life spent in growing from ground-level into the canopy when a gap is formed. Where the average length of life is 300 years only two or three gaps per hectare, each containing one or two young stems, would be sufficient to perpetuate the forest.'

### iii) The Tawny Owls of Wytham Woods, Oxford

Cohesion and self-regulation, irrespective of population density, is clearly demonstrated in a very detailed study of tawny owls, *Strix aluco*, in Wytham Woods [85, 86]. Wytham Woods is surrounded by farmland, and to this extent forms an autonomous system. Within the woods tawny owl pairs hold breeding territories.

The study started in 1947, and lasted 12 y. The winter of 1947 was exceptionally cold in England causing the death of a large proportion of the owl population. Over the following ten years the number of pairs holding a territory gradually increased until an asymptote was reached at a population density approximately double the 1947 value. The recovery was monotonic, the actual increase each year showing little relationship to the availability of food or the vicissitudes of the climate. Population density was evidently regulated by the number of territories established, territories gradually contracting in size as the number of breeding pairs increased. Young birds, marked as nestlings, unable to establish a territory, were frequently found dead outside the study area, apparently having died from starvation.

### iv) Cave Fishes

On the African continent, the biology of the blind cave fish, *Caecobarbus geertsii*, in the caves of Zaire, was investigated by Heuts [87], and in the Southern United States, Poulson [88] investigated blind cave fishes of the genus *Amblyopsis* over a period of several years. *Caecobarbus geertsii* appears to be the only species known to exist in an single-species ecosystem. No other living inhabitants are present in the Zaire caves, the fish being entirely dependent for food on detritus, carried down from the outside world during the six months of the rainy season. For the remainder of the year *Caecobarbus* lives on its reserves. Populations in six separate cave systems were examined. Individuals up to 12 years of age were encountered. Most populations reached a modal size between 50 to 75 mm, standard length; age at modal length varied between five and ten years. In one population only were juvenile *Caecobarbus* found and, anomalously, these juveniles were segregated in a pool near the cave entrance. In all other population no fish of age 0 and 1 were collected.

Similarly, in the caves investigated by Poulson [88], very few juveniles (age 0 and 1) were found, yet over many consecutive years' investigation, the modal age remained at 3 years. Poulson comments:

"It is nearly certain that the census figures of age groups 0 and 1 are erroneously low because census figures of age groups 2 and 3 two years later are higher."

It is inescapable that young fish exist somewhere within the system, yet apparently they do not form part of the active population. A comparable phenomenon has been consistently observed in lake whitefish populations in northern Canada [66]. This is apparently part of the population regulatory process.

## v) Molluscs

A similar age and size-distribution has been observed in the freshwater mollusc, *Anodonta grandis*, in an arctic lake [89]. The modal length of the *Anodonta* was between 75-80 mm and the modal age was 9 y, but Green could find no specimens under 50 mm or under five years of age, despite intensive sampling by divers.

In the deep seas of the Carson Canyon, off Newfoundalnd, Hutchings and Haedrich [90] found a similar paucity of young or small individuals in the molluscs, *Nuculana pernula* and *Yoldia thraciaeformis*. The modal age of both species was between 7 and 9 y. There was little overlap in the size distributions of the two species within one dredge sample, despite the comparable modal age.

## vi) Wisconsin Lowland Forests

A most interesting set of data is presented by Sakai and Sulak [91], which shows the gradual change in a regrowth forest in Northern Wisconsin Lowlands. In 1937 stem diameter showed a negative logarithmic distribution, but this changed gradually to a unimodal distribution by 1980. In this case, three species were involved: northern white cedar, *Thuja occendentalis*, black spruce, *Picea mariana*, and balsam, *Abies balsamea*, all apparently growing in concert.

## vii) Experimental insect Populations

Hassell *et al.* [92], after examining extensive life-table data from a large number of single species insect populations, concluded that: "Most populations show monotonic damping back to equilibrium following disturbance, with only the occasional example of oscillatory damping or some sort of low-order limit cycle." Thomas *et al.* [93] searched for chaos among 27 drosophilid (fruit fly) species at two temperatures and found that single-species laboratory populations invariably exhibit dynamical stability. They suggest that evolutionary pressures acting directly on stability could occur in the form of group selection. In conclusion they add that the pattern of their results "strikingly suggests that indeed the populations are trapped between opposing evolutionary forces." Mueller and Ayala [94] reached similar conclusions although they disputed the contention of Thomas *et al.* that the evolution of such damping mechanisms involved group selection.

## vii) Birds and Mammals

A feature common to all birds and mammals, is the genetic regulation of size which ensures a determinant growth pattern. The tawny owl reaches maturity at the age of one year, therefore the size-frequency distribution is bimodal in spring and early summer, but unimodal during the remainder of the year. This pattern, with variations, is common to all birds and mammals, presumably having evolved from an indeterminant growth pattern such as found in present day fishes. Therefore, it must be presumed that uniformity has important survival value. It seems most likely that genetic control of size is advantageous in land organisms where environmental fluctuation is generally large compared with aquatic habitats. In well-buffered aquatic habitats flexibility allows for 'fine-tuning' of the modal size to the specific environment encoutered.

## The Reference System

The introduction of thermodynamic concepts into ecology can be traced to the influence of Lindeman [95], and in more theoretical vein to Schrödinger [96]. But attempts to relate strict thermodynamic principles to the real world is of more recent origin [97-101]. However, applying the laws of physics to biological systems has not been easy and no generally accepted view has emerged. One feature is evident, application must be done in a flexible manner, for physical theory in this area may not be complete any more than is biological theory. The important aspect is that any additional propositions must not replace established laws, but be additative and complementary.

If ecosystems are to be viewed as thermodynamic entities, the most urgent need is a reference system, against which other systems may be compared, serving the same functions as the *isolated system* in standard thermodynamics. The nucleus of this reference system is to be found in the small arctic lake.

Each of the ecosystems considered is largely isolated from its neighbours. The dominant species in each case assumes a state characterized by high biomass, great age, large size, uniformity in size, a wide variation in age at modal size, and continuity of the configuration. These characteristics are those that would be assumed in a state of *least dissipation* [102, 103, 104]. Least dissipation, or *least entropy production per unit energy stored*, implies the least attainable dissipation in the given conditions (not least in the absolute sense); it is the accumulation of the greatest biomass attainable, relative to the energy input, maintained over the longest possible time period, or the longest attainable time-delay on the energy flow, within the inherent potential of the species concerned and the environmental conditions prevailing.

A state of least dissipation implies uniform size distribution maintained indefinitely. In general, the specific metabolic rate declines with increasing size [105, 106, 107] therefore dissipation, for a given input, is least, relative to the biomass energy accumulated, when all the organisms are of large uniform size. This condition can only be approached but never reached, because of the limited life span of individual components and the need for replacement. Uniform size distribution is equivalent to uniform energy distribution. To achieve the goal of least dissipation, through the regulation of growth and size, species populations must function coherently.

Given the range of taxa from which dominant species may be drawn, it must be assumed that the trend to least dissipation is a universal characteristic of all species, irrespective of their position in the trophic (feeding) hierarchy. In species below the level of dominant, the trend to least dissipation is not so readily apparent on account of predation and herbivory by species higher in the food chain. Being consumed causes more rapid turnover.

The reference system may thus be described as an autonomous system in which the environment is uniform spatially, and temporally regular. Its characteristics may be summarized:

1) The species populations tend to assume a state of least dissipation,
2) The species populations form an organized hierarchial structure based on thermodynamic flows,
3) At the head of the hierarchy is the dominant species. This species, being under least biotic control, most closely approaches its natural limit of (least) dissipation,

It is equally apparent that a trend toward least dissipation cannot be the only factor in ecosystem structuring. There must be a counter trend to that of least dissipation, otherwise energy flow would decline and eventually cease. Thus it is postulated that the conditions of existence necessitate a trend tending to accelerate energy flow.

Heterotrophy is the consumption of one organism by another. The transformation of food into tissue of the consumer demands high expenditure of energy [108]. Therefore, the greater the number of energy transformations, the greater the rate of energy dissipation. This being the case, it is evident from the general evolutionary trend to greater diversity [30, 31, 109, 110], that dissipation per unit of tissue supported has increased over evolutionary time.

Combining the above evidence, it is apparent that uniformity and symmetrical interaction induce least dissipation; increased dissipation results from asymmetrical interaction and increase diversity. The limit of asymmetrical interaction is the consumption of one species by another.

The general statement may, therefore, be made that ecosystem structure is a function of two antagonistic trends; one toward a symmetrical state resulting in least dissipation, and the other toward a state of maximum attainable dissipation. This reduces to an antagonism between homogeneity and heterogeneity. If the system is to maintain continuity, the trend to homogeneity must dominate system behaviour in the short-term (ecological time), but in the long-term (evolutionary time) the trend to heterogeneity dominates system behaviour: hence diversity increases over evolutionary time. This dynamic state may be compared with that of the solar system, in which the planets maintain their position during the 'short-term' by cyclic motion around the sun, but in the long-term they move imperceptibly toward a state of maximum entropy.

At the most fundamental level, ecosystem structure thus reflects a state of antagonism between the basic characteristic of energy to distribute itself as uniformly as possible throughout a system (a second law effect), and the asymmetry contingent on to the maintenance of energy flow. This antagonism is evident only close to thermodynamic equilibrium.

The ecosystem stationary state, or climax, is attained when the two antagonistic trends reach near-equality. This state is therefore a 'minimax' condition in which dissipation reaches the minimum value attainable in the short-term, but this value is also the maximum that has been attained over the long-term. Given these relationships an ecosystem forms an ordered hierarchy based on the individual capacities of the species present to acquire and store energy. The dominant species most closely approaches the limit of its inherent capacity to reach a state of least dissipation, it therefore contributes the major component to the stability of the system. Other species move *toward* a state of least attainable dissipation, but are maintained at a distance from their potential goal by predation and herbivory. A necessary condition of existence is that the ecosystem as a whole also move toward a state of least dissipation.

Given these premises, it is possible to develop a thesis consonant with accepted physical principles.

### The Hypothesis

"We are to admit no more causes of natural things, than such are both true and sufficient to explain their appearances'

*(Isaac Newton, 1686* [111])

It is postulated that ecosystems originated through constraints on the thermodynamic flows arising from the Earth's free energy gradient whose source is the sun and whose sink is the universe.

Following its formation some 4.5 billion years ago, the Earth cooled proceeding toward a state of thermodynamic equilibrium. Near equilibrium spontaneous cyclic events occur. Resonant amplification of these spontaneous fluctuations initated cyclic energy storage processes [99]. Interference between the cyclic energy flows inducing time-delay and the rectilinear flows tending to accelerate energy flow created a state of least dissipation [112]. The energy stored in individual cyclic events tended to distribute itself uniformly throughout the system through homeokinesis [113]. This trend to uniformity was counteracted by the asymmetry essential for the maintenance of flow. Cyclic flows are reversible, or symmetric with time, rectilinear flows are irreversible or asymmetric with time. Flows are accelerated over time under the influence of the Principle of Least Action. Evolution emerges from the interaction of these two principles; the directionality of evolution is caused by the irreversible acceleration of energy flow (Dollo's Law) [114].

This is the most economical explanation for the origin of living things, but like a simple mathematical function, great complexity may develop rapidly when the expression is expanded.

In accordance with these principles the hypothesis may be summarized:

1. The biosystem forms a single closed thermodynamic system within the boundaries of the biosphere.

2.  Initially, spontaneous reversible interactions initiated by quantum fluctuations occurring in the vicinity of thermodynamic equilibrium, were amplified by the resonant effect of an intermittent input of free energy. Energy was thus stored (time delayed) in cyclic processes reversible with time.

3.  These cyclic processes interferred with the rectilinear, irreversible, dissipation of energy causing the system to assume a state of least dissipation, or least entropy production.

4.  Dynamic systems are governed by the Principle of Least Action. Application of this principle induces a constant trend to acceleration of the energy flow.

5.  Least action is the inverse of least dissipation (i.e. least dissipation = most action). Therefore, the ecosystem is structured on the antagonism between trends to least action and most action (least dissipation). For continuity, the trend to most action must exceed or equal the trend to least action in the short-term (ecological time).

6.  In a fluid medium, the cyclic energy storage units induced by resonance (*atomisms or proto-organisms*), interacted and, through homeokinesis, achieved uniform energy distribution.

7.  The resonant amplification of cyclic processes near thermodynamic equilibrium was initiated by signals in the environment.

8.  Environmental signals are created by the intermittent energy flow from the sun, caused by the Earth's rotations, combined with parallel flows of materials [115].

9.  In the initial state of symmetry, uniformly charged proto-organisms functioned as a *tabula rasa* for the identification of signals.

10.  Species were formed by the identification and encoding of specific environmental signal series. They are thus 'hard copy' containing environmental information.

11.  The encoding of signals demands that work be done, therefore, an increase in the number of signals encoded demands an increase in work done. The necessary increase in power is obtained from an increase in energy flux, stimulated by the principle of least action.

12.  Under the stimulus of the principle of least action the ecosystem explores to the full the signal system of the environment.

13.  Individual species tend toward a state of most action. Interaction between species tends to reduce the action.

14.  Over evolutionary time, as diversity increases, the average interactional differential declines. In certain cases the interactional differential declines to near-zero giving rise to symmetrical interaction between species. This allows mutualism and synergy to develop.

15.  When an individual species increases its action it becomes more heavily damped, thus able to respond to a wider range of signals. Responding to a wider range of signals allows the species to encroach on the signal series of neighbouring species. This is the origin of competition.

16.  Distortion of the initial state of symmetry results in the formation of a hierarchy. This hierarchy is formed by the interaction between uniformity and sequence. Uniformity is maintained within the individual species, sequence

results from the varying capacities of the species complement to acquire and temporarily store energy.

17. The hierarchy can persist only in a state of distorted symmetry. That is, overall near-symmetry must be maintained if the hierarchy is to continue is existence.

18. Homeostasis results from the short-term ascendancy of the trend to most action. Le Chatelier's Law ensures that the forces inducing most and least action eventually approach equality at the stationary state or climax.

19. Out of the slightly asymmetrical interaction between the trends to least and most action, emerges self-organization and evolution. Evolution weaves an erratic course, either toward increased action in the individual species (increased size, increased mean life-span, etc.) or toward decreased action with respect to the system (increased diversity and increased complexity). These two processes are in a state of continual interaction. Over the short-term, the trend to most action dominates the system behaviour; over the long-term the system behaviour is dominated by the trend to decreasing action. Dominance of the long-term trend to faster energy flow, ensures that the overall directionality of evolution is toward increased diversity and increased complexity.

## Amplification

### The Theorem of Least Dissipation

When a system is prevented from going to equilibrium by the boundary conditions, ". . . . it does the next best thing: it goes to a state of minimum entropy production–that is, to a state as close to equilibrium as possible" [104 p 139]. Close to equilibrium, the entropy production itself functions as a potential [22]. "Far from equilibrium," Prigogine and Stengers [104 p 140] continue: "the system may still evolve to some steady state, but in general this state can no longer be characterized in terms of some suitably chosen potential (such as entropy production for near-equibrium states)".

The theorem of least dissipation states that the entropy production is a Lyapounov function in the strictly linear region around equilibrium. "If the system is perturbed the S [entropy] production will increase, but the system reacts by coming back to the minimum value of the S production"

Given the structure of the reference ecosystem and its behaviour following perturbation, it is evident that, contrary to Prigogine's general conclusion, ecosystems obey the theorem of least dissipation, although far from equilibrium.

The rate of dissipation at the stationary state can be measured in relative terms by the production $(P)$ necessary to maintain a given biomass $(B)$. The smaller the Production/Biomass ratio $(P/B)$ the less the dissipation. As MacArthur (in Leigh [116] has shown, the $P/B$ ratio is dimensionally equivalent to the reciprocal of the mean life-span. The Biomass Accumulation ratio $(B/P)$ [117] is therefore equivalent to the mean life-span, or turnover time. In absolute terms, the energy accumulation should be measured as a function of energy in the biomass (biomass x energy density) and time (mean life-span). The dimensions of this quantity are joule-seconds in the SI system of units.

### The Principle of Least Action

The general principle govering the trajectory of a dynamic system is that it follows the path of least action [118, 119, 120]. The principle of least action, had its origins in the principle of "least time" originally enunciated by de Fermat in the seventeenth century. de Fermat concluded that the passage of a light ray through a series of transparent media of different refractive indices, could be described by the path of "least time," not necessarily the shortest distance. This theorem was generalized by de Maupertuis, who developed the principle of "least action" that "nature acts in such a way as to minimise the quantity he called 'action'" [120]. Lagrange discovered that he could characterize the true path by an "action" whose value was insensitive to small alterations in the path; that is, the action remains stationary under small changes in path. Action in physics is an abstract quantity determined by the difference between the kinetic energy and the potential energy integrated over the time interval of the event under consideration. In practice action may be thought of as twice the kinetic energy of the system multiplied by the time interval between the initial and final position under study.

The basic equation connecting the physical and the biological world thus appears to be expressed by Heisenberg's uncertainty principle phrased in terms of energy and time:

$$\varDelta T \varDelta E \geq \hbar$$

where $\varDelta T$ is a change in the time of an event, $\varDelta E$ is a change in the energy of the system, and $\hbar$ is Planck's constant, which is the quantum of action. Thus very small time intervals imply very high energies if the product is to remain near $\hbar$.

In biology, action in a system at the stationary state, may be thought of as the difference between the production (kinetic energy) and the biomass (potential energy), integrated over a specific period of observation. It may be considered as the average energy (in joules) in the biomass, times the mean life-span of the species (or ecosystem). Biological action is therefore measured in joule-seconds.

The principle of least action indicates that the energy flux (the rate at which energy passes across unit area), relative to the energy accumulated in the biomass (biomass x energy density), in a fluid, changeable, energy-accumulating system will tend to increase. Least action therefore expresses the opposite trend to that expressed by least dissipation. That is, least dissipation is equivalent to most action. Therefore, within an ecosystem, the trend to most action is in permanent conflict with the trend to least action.

Davies [119] states that the principle of least action is fundamental to understanding the sub-atomic quantum world. He developed a instructive analogy depicting the behaviour of sub-atomic particles under the influence of the law of least action. Imagine, he states, people walking through a city park. The majority will take the direct path across the park from the entrance to the exit, but others of higher energy will tend to explore the park to its limits. The larger and heavier the particle the closer it stays to the direct path. Children, Davies suggests, being lighter and smaller, tend to follow less direct trajectories. This has close correspondence with energy flow in the ecosystem. The main pathway through the ecosystem is the path of most action. At the same time it is maintained as close as possible to the path of least action. The main energy pathway through the system, from assimilation to dissipation as heat, terminates in the largest, heaviest, and longest living organisms (and most action). The rarer species explore the full possibilities of the park, but at a higher level of energy expenditure, or higher power.

The non-equilibrium system is thus structured on two conflicting principles, one, operating in the short-term, tending to most action, a state of least work, and minimum entropy production, the other, inducing more rapid energy flow (hence having the potential for increased work) over the long-term. This ensures a state of least work in the present but a gradual increase in power over time. In a fluid system capable of change this slightly asymmetrical antagonism provides i) a mechanism for short-term homeostasis and ii) a mechanism for self-organization.

### Resonance

Resonance occurs when an object is subjected to vibrations in the vicinity of its natural frequency of oscillation. Resonance thus delays the dissipation of energy from the initiating oscillator. In the simplest case, for example the child's swing, oscillation is caused by the pulse of the parental hand at the natural frequency of the swing's motion. Energy is stored by the continual transformation of kinetic to potential energy and vice versa. At the top of the swing the energy is all in potential form, as the passes through the vertical position all the energy is kinetic.

If the receiving resonator is damped, as by the introduction of a resistance in an electrical circuit, it becomes responsive to a wider range of frequency.

### The Origin of Ecosystems

To exist within the context of the physical processes outlined above it is necessary to postulate that living systems first developed in conditions close to thermodynamic equilibrium. Near equilibrium spontaneous reversible (cyclic) processes occur. If such spontaneous processes are amplified by an energy input the action will increase, increasing the life-time of the cycle. One of the electrons in the responsive atom assumes a wider orbit, remaining in this excited condition for $10^{-7}$ to $10^{-8}$ s [121] before returning to the ground state with the emission of a photon. In the excited state the atom is more reactive, therefore able to form more complex compounds.

The excited entities, or atomisms, thus formed, exist at the intersection of cyclic and rectilinear energy transport processes (reversible and irreversible energy flows). These transport processes interfere with each other so that the system goes to a state of least dissipation [112].

Periodic stimulation causes the receptive atomism to resonate, inducing higher energy levels and longer time-delays (increased action). The longer time-delays induce increased damping, enabling the organism to respond to a wider range of signals. Identical atomisms will resonate in synchrony [122], but destructive levels of oscillation will be inhibited by the damping effect of the increase in action. Increased action extends the time delay and increases the damping moment, least action tends to return the atomism to the ground state as rapidly as possible with the emission of a photon (this occurs when the biological pathway is short-circuited and the accumulated energy is burned).

The extended time-delay allows work to be done, thus processes may develop within the extended cycle that require internal work for their execution. Such processes increase dissipation (i.e. they are "dissipative processes"). Thus the capacity for internal work allows complex "dissipative structures" to develop. Dissipative structures develop through the antagonism between trends, first to most action and energy accumulation, and then least action. The near-equality of the trends to deceleration and acceleration of the energy flow in a labile system leads to maximum power output; that is, when the rate of energy expenditure is at the mid-point between zero dissipation and instantaneous dissipation, and the efficiency (output/input) is 50 percent [123].

Thus a homeostatic mechanism is formed tending to return the system to a state of minimum flux and minimum work, as a result of the short-term dominance of the trend to most action. However, because this situation is really a stabilization of flow, not the stabilization of a specific state, the term *homeorhesis* is to be preferred to homeostasis [124].

## Signal Identification

The resonance of susceptible atomisms through periodic stimulation, implies the existence of an environmental signal or signal series. The initial resonant atomisms, or proto-organisms, identified and encoded the primary environmental signal. 'Information' was extracted from the environment. The atomisms resonated in synchrony distributing the temporarily stored energy uniformly among themselves. The uniform distribution of energy among charged atomisms has been described by Soodak and Iberall [113] under the doctrine of homeokinesis. Atomisms equipartition the energy through an interactive process. "Such complex atomisms do not equipartition the energy per collisional cycle, but instead internally time delay, process, and transform collisional inputs, generally using many fluidlike dissipative mobile steps" [113]. In that work is done in the processing and transformation of collisional inputs, homekinesis is a dissipative process. Because the system tends to a state of least work, or least dissipation, and less work is necessary near the symmetrical state, the system tends to uniformity. Least dissipation is thus contingent on the operation of the homeokinetic mechanism.

The initial state of symmetry induced and maintained by homeokinesis was, in effect, a *tabula rasa* made up of uniformly charged atomisms, equivalent to a set of indeterminate symbols. According to Polanyi [125] a state of indeterminacy must exist among symbols if they are to function as a code: "It is this physical indeterminacy of the sequence that produces the improbability of occurrence of any particular sequence and thereby enables it to have a meaning–a meaning that has a mathematically determinate information content equal to the numerical improbability of the arrangement." The uniform atomisms were thus in the requisite state for coding information.

The primitive system was thus capable of identifying specific signals within the environmental signal series. Environmental signals are periodic fluctuations in the parallel flows of energy and materials in the biosystem [115]. When this occurred the symmetry was *distorted*, not broken [126,127]. To continue to function, overall symmetry had to be maintained. As Iberall *et al.* [128] conclude, a state of *equipollence* (equal power) must be maintained. Distorting the symmetry, through an increase in signals identified, demands an increase in power [129]: the greater the number of signals identified, the greater the power output necessary. The system thus tends to identify as many signals as possible in compliance with the long-term trend to maximum power. A "minimax" condition is formed in which, in the short term, the system assumes a state of least work (as do all physical systems) but at the same time this is a condition of most work when viewed from the long-term perspective.

### Application of Physical Principles in the Living World

The stationary state, in which these principles are most evident, exists at the present time only in isolated parts of the world. Other systems move continually toward a stationary state but are continuously displaced by further disturbance, either from natural environmental variability or man's direct intervention.

The importance of uniformity and cohesion within the individual species population is now seen as a necessary condition for existence. Each species population with the reference lake, functions as a "dissipative structure" in that it tends to assume a state of least attainable dissipation, commensurate with the environmental conditions, its position in the hierarchy and its genetic make-up. Juveniles, with higher specific metabolic metabolic rates than the larger adults will tend to increase dissipation, therefore must be maintained at the minimum level necessary for replacement purposes.

It is thus postulated that there are two fundamental but antagonistic organizing principles determining ecosystem structure:

1) a trend to most action, expressed by individual species populations in cohesion and homogeneity, and
2) a trend to least action through increasing diversity and increasing complexity.

The two trends will develop under the controlling pattern of energy input, the individual histories of the component species [100], and the environmental history. If the system is to maintain continuity the trend to most action must dominate the system behaviour in the short term, whereas the trend to increasing dissipation dominates the system behaviour in the long-term. Out of this antagonism emerges evolution.

## Evolution and Diversity

The central theme of neo-Darwinian evolution is natural selection, the survival of those species fittest to survive, with fitness defined through the process of differential reproductive success. No mechanism has been postulated to account for this process, apart from the inherent nature of the process itself. In fact, one of the foremost proponents of the 'Modern Synthesis' maintains that this is not necessary [28, 130]. The hypothesis outlined above provides the essential mechanism.

In the short-term, individual species, as well as the system as a whole, tend to assume a stationary state of most action and least work, but an idealized stationary state as postulated in the reference system, will seldom be attained in the real world because of environmental change and variability. For much of its existence an ecosystem will be *stationary-state-seeking*. Over evolutionary time-action declines, through increased diversity and increased complexity, necessitating an increase in internal work and an increase in dissipation.

To maintain a position in the ecosystem hierarchy a species population must tend continually to increase its action through increased abundance, increased uniformity, increased size, increased mean life-span, and increased energy acquisition.

The trends to least and most action may be regarded as *heuristics*. In computer terminology a heuristic is defined as a "method of behaving which will end towards a goal which cannot be precisely specified because we know *what* it is but not *where it is* [131, p 69]. Thus the instruction "keep going up" would be the heuristic for climbing an unknown mountain. "Heuristics prescribe general rules for reaching general goals." The individual species thus obeys the heuristic "increase action" whenever opportunity arises; likewise, the system as a whole obeys the heuristic "decrease action" whenever possible. These heuristics are not symmetrical. The trend to most action is reversible in that cyclic activity is symmetrical with time, whereas the trend to least action is irreversible. This irreversibility functions as a pawl in the evolutionary ratchet, whereby the system is gradually "winched-up" to greater diversity and greater complexity.

This necessitates a somewhat different view of fitness from the conventional one. The fitness of a species is a measure of its ability to hold a position in the ecosystem hierarchy over an extended period of time. That is, fitness is the equivalent of action ( = energy density × biomass × time). As Van Valen [132] states, fitness is the capacity to control resources. An increase in fitness improves the position of a species within the hierarchy and this will be reflected in differential reproductive success. An increase in action increases the damping moment of a species allowing it to respond to a wider range of signals. At the same time increased action will increase the retention of nutrients, preventing their utilization by competitors. Through encroachment on neighbouring signals, a species improves its fitness with respect to its neighbours. The dominant species at the head of the hierarcy, as McNaughton and Wolf [133] point out, is the species capable of utilizing the greatest range of resources.

There is thus continual tension within the ecosystem between individual species tending to increase their action and the system tending to accelerate energy flow.

Driven by this slightly asymmetrical process, evolution weaves an erratic course. Evolutionary lines may show an increase in the size of inviduals (Cope's rule) [134], an increase in life-span, energy density, or complexity. An improvement in energy acquisitive capacity will demand an increase in complexity and a comensurate increase in energy flux. An increase in complexity will therefore be effective (i.e. will be 'selected') only if the integrated position of acquisition and storage is improved. This ensures that any increase in complexity beyond the immediate need will be minimal [135].

Increased complexity is a two part programme. The first part is the necessary acquisition of energy through an increase in action and then the reworking of this additional energy within the continuously accelerating energy stream. The simplest and most economical way to increase biomass is to replicate similar parts; then to meet the need of maintaining position these parts assume specialized functions. The over-riding trend to increased dissipation gradually fines down or 'streamlines' any increase in energy accumulated.

Central to the thesis is the proposition that the energy flux through the biosystem, relative to the biomass supported, increases over evolutionary time. Evidence for such a proposition is, necessarily, indirect. If the 'information' in the biosystem has increased over evolutionary time, as is evident from the overall increase in diversity [30, 31, 109], then the additional power necessary to support this increase must come from somewhere. An increase in diversity among plants doubtless results from the exploitation of an increasing range of energy sources (e.g. adaptation to lower light levels) augmented by the 25% increase in solar power over the history of the Earth [136]. However, it seems highly probable that much of the diversity in plants is attributable to animal activity. Plants are consumed, preventing them from reaching a stationary state, thereby ensuring that they maintain a dynamic mode which increases the opportunity for new pathways to arise.

Thus, Stanley [137, 138] attributed the sudden increase in diversity in the Precambrian and Cambrian Periods to the origin and development of heterotrophy: the consumption of one organism by another. Stanley concluded that grazing by the newly evolved heterotrophs on the formerly dominant stromatolites (communities of algae and bacteria which had dominated the ecological scene for the previous two billion years) [139] promoted the diversity of both plants and animals.

This grazing hypothesis is substantiated by an ecological counterpart recorded by Jacobs [140] from Lake Nakuru in Kenya. Lake Nakuru is a soda lake having initially a very limited biota.

"Prior to 1961, there were essentially only one or two species of algae, one copepod, one rotifer, corixids, notonectids and some 500,000 flamingoes belonging to virtually one species."

Ostensibly to control mosquito larvae, the cichlid fish, *Tilapia grahami*, was introduced into Lake Nakuru in 1962. The fish flourished, feeding on the algae as mosquito larvae were relative scarce, and although reaching only small size, a very high density was attained in the absence of predators. However, before long predators, in the form of fishing birds, moved in. About 30 species of fish-eating birds, including pelicans, cormorants, herons, egrets, grebes, terns and fisheagles, formerly absent from the region, congregated to exploit the new and abundant resource. As in the Cambrian Period a single key species was sufficient to broach the reservoir of accumulated energy, converting it to a more dynamic state.

The biosystem may be regarded as a great communications network [141], in which the primary producers explore all avenues to pull in more signals and increase assimilation. At the same time, primary producers, unable to eliminate all regular fluctuation, function, themselves, as signals capable of being identified by higher level resonators. A highly damped resonator will assimilate the energy in a signal less effeciently than a lightly damped one closely synchronized with the signal. Thus, again, individual species in the system are driven in two directions. One direction is toward increased damping and less efficient assimilation; the other is toward greater specialization and more efficient assimilation. The outcome is that the environmental signals are identified to the greatest extent possible and then reshaped within the energy stream. The result, as Southwood [51, 142] suggests, is that the habitat functions as a 'templet' around which species assemble.

An environment of high energy input in which the signal series exhibits little variability within and between years will be the one most readily dissected. The initial homogeneity of the input must, however, be broken up into a series of microscopic fluctuations in time and space, if diversity is to increase. This is evident in the terrestrial environment of the wet tropical regions. There is a regular periodicity in the temperature, sunlight and moisture, providing continuously high energy availability and allowing assimilation throughout most of the year. The dissection of this relatively uniform signal began when terrestrial plants evolved and developed structure. Once structure developed diversity became self-augmenting as the massive energy flow became fragmented into innumerable minor channels each capable of being identified by a specific organism. Diversity then increased over long periods of time until an asymptote was reached.

Environmental homogeneity or predictability [43, 44], by itself, is insufficient to account for diversity. Few environments are more stable, or more predictable than the mid-waters of the Arctic Ocean [143], yet diversity in these waters is extremely low. Similarly in the open oceans of the tropics the environment is extremely uniform, the water-column is stable and nutrients quickly leave the photic zone. In the absence of a suitable substratum on which biological structure can develop, there is little possibility of high diversity developing. Consequently, diversity in the oceans, develops to its greatest extent on coral reefs where structure provides for increased environmental heterogeneity.

From the tropics to the polar regions, climatic variability increases and energy availability decreases commensurate with the decrease in diversity. The decrease in diversity results from the increasing difficulty in identifying signals of high variability, and the need to accumulate sufficient reserves top meet variable conditions. It is evident also that the specific metabolic rate which is temperature dependent in all species except birds and mammals, will be greater in the tropics than at higher latitudes. Other things being equal, the high metabolic rate in the tropics will stimulate faster energy flow than in temperate regions. This inherently higher energy flow, combined with all other factors ensures, a greater rate of acceleration of energy flow in the wet tropical regions compared with other regions. This is supported by the conclusion of Stehli *et al.* [144] that evolutionary rates are higher in the tropics than elsewhere.

The relationship between productivity and diversity is not a simple one. Time and the temporal structure of the environment are of great importance. Relatively few studies have shown conclusively that species diversity increases with productivity. In one of the few recorded cases Brown [145] showed that in comparable desert habitats involving rodent species, the more productive system was the more diverse. However, Abramsky [146] found that both productivity and structure, in comparable desert habitats in Israel, are important factor in high diversity in such conditions.

In many systems high productivity is marked by relatively low diversity, a condition particularly apparent in areas of ocean upwelling, which are amongst the world's most productive ecosystems. High productivity and low diversity is favoured by a highly pulsed energy input combined with high total energy. This combination is the basis of high agricultural production.

The nature of the system boundaries also effects the development of diversity. In an autonomous system biotic constraints become more severe so that the effect of dominance will be most noticeable in the reduction in diversity. As Connell [147] showed, diversity increases in systems subjected to a level of perturbation just sufficient to eliminate the dominant species.

Thus, diversity increases to the greatest extent possible within the physical and biotic constraints of the environment. High diversity is dependent on environmental homogeneity and high energy availability at the macroscopic level, but only when there is heterogeneity at the microscopic level. Heterogeneity may be either in time or space. Greatest diversity occurs where these conditions have obtained for a long time.

### Hierarchy Formation

Within the biosphere boundaries a hierarchy develops that controls and regulates the energy flow. In recent times the conceptual basis and structure of the hierarchy has attracted considerable attention [148–154], but no satisfactory causative mechanism has been developed.

The hierarchy, it is postulated, is a dynamic dissipative structure, maintained by work done. It is the current configuration of the initially distorted symmetry arising from the conflict between the basic properties of order: sequence and uniformity (rank and row). Thus a hierarchy consists of rows or uniform entities arranged in some pre-determined sequence. The group of uniform items, Koestler refers to as a *holon* which he defines as a semi-autonomous sub-whole equipped with self-regulatory devices [148, p 97]. Within a holon, energy is distributed uniformly among members of a chohesive group. Sequence arises from the energy flow, uniformity from the inherent trend of energy to assume a symmetrical state. The hierarchy therefore functions as a graded constraint on energy flow. Sequence is determined by the differing abilities of the holons to acquire and conserve energy; the one with the greatest capacity occupies the head of the hierarchy, the remainder being arranged in descending rank.

Within the world system various sub-hierarchies, or ecosystems are formed. Again, within each holon uniformity may break down with the formation of sub-hierarchies. Thus some species exhibit great uniformity, such as the herring school, others, such as the wolf-pack develop a hierarchical system, responsive to a wide range of signals to which a graded response can be made. At the base of the hierarchy are the many small organisms, working hard to maintain their position; and at the apex is a relatively small number of large organisms doing relatively less work to maintain unit biomass. Waddington [155] states "Bastin . . . argued that the only logical way in which it is possible to discriminate a number of activities into a hierarchy is by considering their reaction times, a higher level in the hierarchy having a much larger reaction time than a level classified as lower." The rate of energy flow is gradually reduced along the food-chain. For a hierarchy to remain in existence each holon must maintain its discreteness. Variation must therefore be discontinuous.

Thus the ecosystem hierarchy develops from the individual species each attempting to attain a state of most action possible through cyclic, reversible flows, against the irreversible trend to increasing energy flow, induced by the principle of least action. The ultimate pattern of flow is worked out among the assembled species as they jostle and compete for resources during the course of succession.

### Succession

Succession is the primary process whereby organisms assemble and establish a hierarchy within the boundaries of the system and constraints of the environmental signal series.

The concept of succession has been developed largely by plant ecologists to express the changes that take place following the initiation of conditions suitable for plant growth and the ultimate establishment of stable, climax vegetation. Primary succession develops on new sites, such as, characteristically, the conversion of a shallow lake or river delta, to forest, through erosion, sedimentation and soil formation, or the colonization of a moraine in the face of a retreating glacier. Secondary succession is the re-vegetation of a site formerly occupied by the climax biota. This is characterized by the regrowth of vegetation in an area cleared by a falling tree, or the regrowth of forest on land formerly cleared for agriculture. Both processes follow a similar trajectory.

Understanding succession is of great practical importance, because for most purposes, existence takes place within a successional world.

McIntosh [6] states:

"Succession is one of the oldest, most basic, yet still in some ways, the most confounded of ecological concepts . . . The apparent intractability and continued contradictions of the successional question after many decades of study, lead to the suspicion that there is more involved than straight forward, matter–of–fact, scientific consideration."

According to McIntosh [6], Clements [15] and Odum [18, 156] converge in their description of succession as an orderly, predictable, unidirectional process which results in modification of, and control over, the physical environment culminating in a mature (or climax) ecosystem, largely governed by climate. Gleason [157] opposed these views. He describes succession:

" . . . as an extraordinarily mobile phenomenon whose processes are not to be stated as fixed laws, but only as general principles of exceedingly broad nature and whose results need not, and frequently do not, ensue in any definitely predictable way."

Whittaker [158] in a detailed review of the climax concept, comes down fairly heavily on the side of Gleason, stating:

"Succession may thus be thought to occur, not as a series of distinct steps, but as a highly variable and irregular change of populations through time, lacking orderliness or uniformity in detail though marked by certain fairly uniform overall tendencies."

More specifically, one of the most general characteristics of the successional process was originally described by Clements [49]:

". . . the number of species is small in the initial stages, it attains its maximum number in intermediate stages; and again decreases in the ultimate formation, on account of the dominance of a few species."

The initial trend toward increasing diversity is not necessarily a monotonic progression: periods of increasing diversity may be interspersed with relatively stable configurations in which certain species temporarily dominate before giving way to the next phase of increasing diversity [159, 160].

This highly significant pattern has subsequently been confirmed by many observers [17, 18, 117, 158, 161–167]; despite this evidence McNaughton and Wolf [130] question the generality of the pattern.

A further important generalization is the gradual replacement over the course of succession of relatively short lived species by species of longer life-span. This is the replacement of $r$-selected species by $K$-selected species. The general trend in energy flow during succession is thus characterized by a complex relationship in which energy flow increases with increasing diversity, yet at the same time is constrained by the ascendancy of the trend to most action. Only in the terminal stages, when the dominant species is ascendant and diversity declines, is the system trajectory evident. In these final stages the trend to most action is evident in the generally recognized decline in P/B [18, 169].

Pickett [170] suggests that succession is a "complex gradient of decreasing physical stress."

Given sufficient time and continuity of environmental conditions, the ecosystem assumes a relatively stationary state or *climax*. In the final form "the community tends to damp environmental fluctuation, to stabilize its microclimate and environmental chemistry" [117]. Clements [49], in fact, regarded stabilization as essentially synonymous with succession. The overall trend, therefore, is toward a state of *least fluctuation*. The stability of the climax is largely dependent on the characteristics of a relatively few dominant species [117, 171, 172]. However, Jones [84] suggests most pertinently:

"It is conceivable that 'climax forest' is a concept only, never existing in practice, either because of catastrophic initiation of fresh seres, or because of the time lag—necessarily great where long-lived trees are concerned—in the adjustment of the vegetation cover to an ever-changing environment."

Even in the idealized state the climax is far from being a static condition: it is made up of many cyclic events taking place within different time frames. The succession of deaths that inevitably occurs in the dominant tree species of a forest creates a mosaic of patches at different successional stages in which, as Cooper [173] stated:

"The forest as a whole remains the same, the changes in the various parts balancing each other."

Over long time-spans the characteristics of the vegetation may change in a cyclic manner as a result of natural events, as described by Watt [174] and illustrated by cyclic changes in Tregaron Bog in Ireland, where heather and spagnum moss replaced each other at intervals of about 20–50 years. Similarly it is now recognized that naturally occurring, long-term cycles may be induced by fire as in many North American forests, or by other recurrent phenomena such as hurricanes [98, 166, 175–177].

Within these major cycles are cycles on a much shorter time scale. In developing their hierarchical concept Urban *et al.* [154] state:

"Events at a given level have characteristic natural frequencies and, typically, a corresponding spatial scale. . . . Higher levels may be larger than, slower than, constrain (control), or contain lower levels. . . . . . A component at a given level of a hierarchy experiences as variables only those patterns that are similarly scaled in rate, as well as size. By comparison lower-level dynamics are so fast that they are experienced as average values; higher level dynamics are too slow to be experienced as variables."

Odum [18] and Dunbar [20] suggest that succession is a short-term reflection of evolution. Odum [18] states:

"In a word, the "strategy" of succession as a short-term process is basically the same as the "strategy" of long-term evolutionary development of the biosphere—namely, increased control of, or homeostasis with, the physical environment in the sense of achieving maximum protection from its perturbations."

This is undoubtedly true, but great care must be exercised in distinguishing between succession and evolution in this context for they are not symmetrical terms. Succession is the establishment of a hierarchy of (reversible) cyclic processes maintaining overall continuity and a state of most action. This is opposed by a countervailing (irreversible) trend to least action, a process that may be described by the term *abscession*—a moving away. *Evolution emerges out of the conflict between succession and abscession.* Thus evolution represents the sequence of changes that result in the operation of successional processes at increasing energy flow rates. Succession reaches at least a theoretical end-point at the climax, the end-point or goal of abscession is maximum power. The abscessional process is open-ended and irreversible. Out of the antagonism emerges evolution, but the results of the evolutionary process must be such as to maintain the near-symmetry of the climax configuration at all times.

### Energy Flow Within the Ecosystem

In the reference system of the simple autonomous arctic lake, free energy flows from the primary producing organism, through a bacterial or detrital loop (which forms the main energy reservoir [178]), to benthic invertebrates and finally to fish. For an ecosystem to be dynamically stable, action relative to energy input must increase along the food chain. Action is a function of biomass, energy density, and time.

The main energy pathway follows the path of increasing action, but species more energetic than the majority will explore the possibilities of the ecosystem to the full. This accounts for the many apparent aberrations and anomalies and rare species. These species will tend to have a higher Power/Economy ratio [179, 180] than those on the main path.

Each step along the food-chain represents a step toward increasing action. Energy flows against the operation of the principle of least action. Thus the second law is directly confronted and overcome in these dynamic conditions. This flow pattern had its origins close to thermodynamic equilibrium in the interference between cyclic and rectilinear energy transport processes. In these circumstances, entropy production functions as a potential. Over the course of evolutionary time, and through increased energy flux, this mechanism has been amplified. Maintenance of this flow pattern is possible only if the system remains in a state of distorted symmetry.

The biomass accumulation ratio of each species may be considered as an approximate measure of its *biological temperature* (the energy of the system relative to the entropy production), the parameter determining the direction of energy flow between or within systems [98]. Within an autonomous ecosystem it can be seen that flow is from low biological temperature to high biological temperature; that is, energy must be "pumped" by the Onsager "reciprocal relations pump" from low to high temperature. At the same time the total action at each trophic level may decline resulting in the formation of "trophic pyramids" [178, 181–183].

This position is supported by the several characteristics of ecosystem structure previously identified: i) increase in size along the food chain, ii) increase in energy density, iii) a decline in the $P/B$ ratio. These characteristics appear to be general in ecosystems. Sheldon *et al.* [184] concluded that size increases with certain regularities along the food-chain. These observations have been developed and expanded by Kerr [185], Platt and Silvert [186], Dickie *et al.* [187], and Sprules [188]. As specific metabolic rates decline with an increase in size, it can be concluded that specific metabolic rates decline along the food-chain in accordance with the trend to increasing action.

Comparably, Bonner [189] and Fenchel [190] concluded that the rate of reproduction declines with body size.

For ecosystem stability, Ulanowicz [191] and May [192] concluded, energy density (joules per gram) must increase along the food chain. Hirata and Fukao [193], Hirata [194] and Hirata and Ulanowicz [9] corrected and developed the original thesis. Hirata and Ulanowicz [9] stated, significantly, that the direction of energy flow may be reversed locally; it is the overall direction that is of ultimate importance.

The validity of this proposition has been tested in a small arctic lake and found to be correct [178].

Several authors have concluded that that the P/B ratio tends to decrease along the food-chain [195–197]. A decrease in P/B implies an increase in B/P. Southwood [51] comes close to incorporating all the features in a single expression. Size, he states, is proportional to generation time and this is proportional to the reciprocal of the specific metabolic rate and in turn proportional to the reciprocal of the per capita rate of increase.

In the reference system it is readily apparent that energy flows from the primary producing algae (small, short-life span, low energy density) through organisms of intermediate size and life-span, then, finally, to the fish predator of high energy density, large size and long life span.

Such relatively straight-forward sequences are obscured in many ecosystems. In forest systems no relatively simple chain can be identified. With the emergence of life onto land, the primary producer regained the dominant status formerly held by algae in the stromatolite associations of the early Precambrian, prior to the evolution of heterotrophs. Interestingly, the evolution of land plants is a major evolutionary step in the direction of increasing action. It is as if the primary sequence from algae to fish dominant has been condensed into a single species: the forest tree. Animal species then establish secondary series on the fluctuating production of leaves and flowers of the dominant, in combination with the co-existing shrub and herb vegetation. On grasslands, however, a sequence similar, but shorter, to that of the lake ecosystem has evolved: from primary producer to grazing dominant, with greatest energy reserves in the soil compartment [198]. The terminal predators in such systems do not occupy a position astride the main energy flow, but perform what is essentially a scavenging role.

### Global Energy Flows

The world's ecosystems are open systems, exchanging matter and free energy with their neighbours. Nevertheless, openness does not mean complete loss of integrity. The individuality of ecosystems is a reality although the boundaries may be nested within wider boundaries and these boundaries may fluctuate and be impossible to define. Nonetheless, within the changing boundaries persistent hierarchies develop.

The loss of energy to neighbouring ecosystems may be described as "leakage." Leakage, or energy flow between ecosystems, flows in the opposite direction to energy flow within the ecosystem. That is, energy flows from high biological temperature (high biomass energy relative to entropy production) to low biological temperature (low biomass energy relative to entropy production). On the world scale this is represented by an energy flow from polar systems at high biological temperature toward the tropics. As in the ecosystem, these patterns may be temporarily or locally reversed. Comparably, Margalef [162] suggests that a more mature ecosystem will exploit the less mature one.

Evidence to support this thesis can be gathered from two lines of approach. First, it is necessary to establish that there is a biological temperature gradient, second, that flows in the direction indicated do occur. The faster turnover time in tropical regions compared with polar regions is indicated by the findings of Mann and Brylinsky [199] that the P/B ratio of freshwater lakes declines with increasing latitude. A comparable relationship has been recognized for terrestrial ecosystems. Golley [200] estimated that the B/P ratio for tropical forests has a value of about 3.0 compared with 30.0 for temperate forests as determined by Whittaker [201]. For temperate systems Reiners [202] gives a value of 1.03 for prairie, 6.63 for savanna, and 20.6 for forests. Burger [203] concluded that angiosperms (flowering plants) with their faster life-histories, are better adapted to the pace of life in tropical rain forests than gymnosperms (coniferous trees). Hartshorn [204] states that turnover rates in a tropical forest are high, even for the forest dominants.

The actual energy flows from the Arctic to southerly latitudes can be observed in the migration of birds and mammals to northern latitudes in spring and their return south in the fall. During their stay in northern regions they feed on the relatively abundant summer production, and reproduce. The result is that more animals return south in the fall than make the journey north in spring. This process ensures a net export of energy to southerly regions. Migration patterns develop because of seasonal inversion of biological temperature relationships.

The global pattern of energy flow runs counter to the flow within the ecosystem. The tropical regions, as the regions in which action is least, relative to input, become a "sink" for terrestrial flows. The more or less continuous energy input in tropical regions, maintains a biological temperature gradient that does not change direction throughout the year, hence migration patterns cannot develop. This is substantiated by MacArthur's [205] statement that there is very little bird migration into and out of the wet tropics. The constant biological temperature gradient throughout the year preserves the autonomy of the tropical rainforest regions. Within this autonomy stability develops.

The global flow pattern is comparable with the flow over evolutionary time: geography (space) and time are interchangeable. Geographically, energy flows from the simple arctic system to the tropical regions along a gradient of decreasing biomass relative to flux (although total biomass per unit area increases) and increasing complexity. Similarly, diversity and complexity have increased over evolutionary time, with, presumably, a concomitant decrease in B/P ratio. These trends must remain within the constraints demanded by distorted symmetry. However, when subjected to stresses, either physical or biotic, beyond the capacity of the system to accommodate, symmetry is lost and the system discards a large number of species, reverting to a simpler state before once again resuming the march to increasing diversity [109, 206–210].

## Ecosystem Stability

In that ecosystems maintain continuity, they have an inherent degree of stability imparted by the capacity of each species population to proceed toward a state of most action. This necessitates a degree of internal cohesion. The degree of cohesion is highly variable, and is most evident in the upper ranges of the food-chain (e.g. tawny owls and northern fishes). From the opposite point of view ecosystems are inherently unstable in that they depend for their existence on the continuity and regularity of a signal series. Ultimately, a system is only as stable as the generative signal series on which it is dependent.

It is therefore postulated that any autonomous ecosystem, irrespective of its diversity, will assume a stable state. That is, in the undisturbed state an autonomous ecosystem will assume the most stable state attainable within the constraints of the environment and the species complement. There are two components to this stability, one 'top-down' the other 'bottom up'. Top-down control results from the tendency of each species to assume a state of most action, along a gradient of increasing biological temperature. This process is ultimately controlled by the ecosystem boundaries imposing density dependent effects. However, over-emphasis on self-regulation through density effects has obscured the important inherent control mechanism operating 'far–from–maximum–density.'

Opposing top-down control is "bottom-up" control, the system trend to least action. Least action implies fragmentation of the energy flow (increased diversity and increased complexity). In that the successional trend to most action is superior to the trend to least action, the system moves toward the stationary state in a well-damped manner.

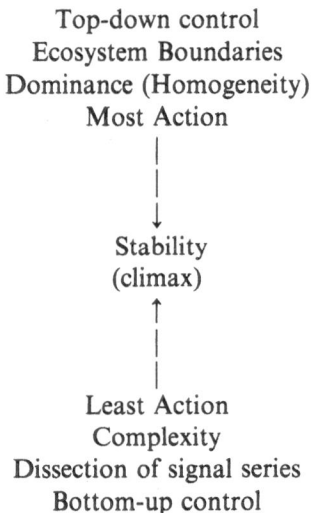

Top-down control
Ecosystem Boundaries
Dominance (Homogeneity)
Most Action

Stability
(climax)

Least Action
Complexity
Dissection of signal series
Bottom-up control

Northcote [211] recognizes the importance of "top-down" control in aquatic ecosystems as well as the existence of "bottom-up" controlling factors, but he does not place the two in opposition as the basic components of ecosystem stability.

Along the main energy pathway the dominant species is the one that attains the "best" integration of energy acquisition and storage. Other species may assume a comparable position at the termination of side-branches in the food chain. Species subjected to being grazed or predated upon at lower levels in the ecosystem hierarchy cycle in their abundances in more direct response to environmental fluctuation. They will thus evolve life-cycles appropriate to their status and the characteristics of the energy available.

## Stability of Ecosystems to External Disturbance

All ecosystems undergo minor disturbance brought about by 'normal' environmental change and variability. It is to the normal environment that ecosystems adjust, given sufficient time without significant change. Occasionally, an ecosystem may be subjected to a disturbance, the like of which, in type or magnitude, it has had no previous experience. Such disturbances may be caused naturally by unusual frost, drought or windstorm. Today, such changes are frequently the result of man's actions such as fishing, and lumbering. The disturbance remains external if only the variables of state are altered [65]; if the disturbance is repeated then it will assume the force of an internal disturbance changing the driving variables (moisture regime, nutrient flow, species complements, etc.).

Based on thermodynamic principles, Hurd *et al.* [212] define stability as the ability of a system to maintain or return to its ground state after external perturbation. The *degree* of stability they characterise as:

i) the amplitude of the deflection from the ground state,
ii) the rapidity of the response to the perturbation,
iii) the rate at which the deflection is damped.

Investigation of small arctic lakes has shown that they meet the criteria of stability, in that after severe disturbance they return to the ground state, given sufficient time [60, 66]. In the cases investigated the time to return was between 8 and 10 years. The degree of stability can be estimated from the criteria of Hurd *et al.* [212]: i) the deflection, at 70% removal of the fish biomass, must be adjudged severe, ii) the response time of about 8 years seems intermediate–to–long for a fish population, although comparative data is scarce, and iii) the deflection was highly damped as indicated by a monotonic trajectory of recovery. Such a population must be adjudged stable but with a comparatively long recovery time. The monotonic recovery process, similar to that observed in the tawny owl population in Wytham Woods (p. 00), and in experimental insect populations (p. 00), indicates the existence of efficient internal, density independent, damping mechanisms.

The severity of the disturbance indicates global stability.

The Arctic charr population in a small lake is evidently stable. It is not, however, particularly resilient, owing to its relatively long time to recovery. Resilience, in effect, is a combination of response time and damping. A resilient species has a rapid response time, and returns to the ground state in a well-damped manner. It is, nevertheless, apparent that these terms are mutually antagonistic. A rapid response time implies a rapid rate of reproduction and fast growth, but a highly damped species will generally have large biomass and be slow growing. A "good" combination of rapid response time and damping moment, are essential for resilience. Every species will have individual resilience characteristics. Species dominant in the hierarchy will tend to be less resilient than those at lower levels. Species fluctuating in abundance will tend to be resilient. Other species may express a high degree of damping but a very long response time.

An example of a stable and persistent fish species in the face of the environment normally experienced, but with very little resilience in the face of new perturbation, is the lake sturgeon, *Acipenser fulvescens* [213, 214]. Most authorities consider that, although initial biomass is high, large-scale fisheries for lake sturgeon are unattainable [214]. Lake sturgeon are known to grow to a maximum size of 90 kg and an age in excess of 100 y. In Lake–of–the–Woods, on the Ontario-Minnesota Boundary a small fishery was maintained (largely for isinglass) through much of the nineteenth century, but the advent of a commercial food fishery around the turn of the century caused yields to increase for a few years, then decline to virtual extinction after about 20 years.

The criteria of Hurd *et al.* [212] have interesting correspondences with the concept of *r*- and *K*-selection [51, 142, 216–217]. The response time of a species is dependent on "*r*", the intrinsic rate of reproduction, which in the more generalized form of the concept is equivalent to the replacement rate. Similarly the damping factor is related to "*K*" the equilibrium constant. Dominant species are considered to be *K*-selected, more rapidly turning over species lower in the food-chain, *r*-selected. Resilience thus depends on a species having a "good" combination of the antagonistic *r*- and *K*-selected characteristics. All species possess an *r*- and a *K*-component, and the ratio $r/K$ is, in fact, homologous with the P/B ratio, which in turn is related to the biological temperature [60].

At the other end of the scale, a highly diverse system is stable locally but unstable globally and of low resilience. The effect of a small disturbance is rapidly disseminated throughout the system, because of the high rate of turnover. However, a severe disruption may require a very long time for rebuilding the complex energy channels formerly existing, hence it cannot be rated globally stable. Land completely cleared of tropical forests may be essentially incapable of recovery [218]. Little is known about the recovery times in diverse systems, but Chevalier [219] quoted by Richards [220] believed that the tropical forest at Angkor Wat in Cambodia, at the site of ancient temples, destroyed some five or six centuries ago, had recovered,but still exhibited certain differences compared with the local virgin forest.

## Stability to Internal Perturbation

An internal perturbation is one that changes the driving variables. Most 'pollution' events are of an internal nature. If an external disturbance is a repeated continuously the driving variables are changed and the perturbation, initially external, assumes the force of an internal one. Internal perturbations include the addition of nutrients or alteration in the timing of nutrient and material flows, or any other change in the established cycle of events. One of the most insidious internal perturbations is the imposition of continual change on the ecosystem. This allows no opportunity for the system to adjust to the prevailing circumstances. It is unable to identify former messages in the changing frame of reference, therefore species will tend to be eliminated indescriminately. A comparable blanket effect would be the application of an environmental poison. Some species will be eliminated or greatly reduced, others, less effected, may improve their relative position in the system.

From the above considerations it is evident that any chemical change will tend to alter the system configuration. Not only is the nature and amount of any addition important, but the timing is of equal significance. For example, an addition of nutrients may either increase or decrease diversity, depending on the timing and frequency of the addition. When nutrients are added over a short time-period, as frequently encountered in 'cultural eutrophication,' an increase in one or few species will be favoured (resulting in a decline in diversity), but when the same amount of nutrient is added over a longer period of time, diversity may increase. Neither effect can be predicted with certainty in any specific context, and the result may be obscured by biotic interactions.

### Unstable Ecosystems

Perhaps the most disruptive of internal perturbations is the introduction of new species. This has been particularly devastating when previously isolated ecosystems, such as tropical islands [70] or lakes [221], have been the site of the new introductions. Frequently, introductions have been facilitated by other disturbances, particularly clearing of the initial vegetation. Isolated systems must be considered incipiently unstable in the evolutionary context. It seems that evolution in isolated ecosystems does not proceed as rapidly as in continental areas, so that they are unable to withstand competition from more complex organisms adapted to faster energy flow rates.

Instability in ecosystems is generally associated with a low degree of autonomy. Unstable ecosystems may exhibit either low or high productivity. This is evident in a comparison of two examples of the world's unstable systems: the small mammals of the tundra, a system of low productivity, and the highly productive marine ecosystem off the coast of Peru. When conditions are appropriate, the Peruvian system supports an immense fishery for anchoveta, *Engraulis ringens* [222–223]. Seasonal upwelling of deep ocean water creates an algal bloom resulting in a relatively short food chain, leading to the anchoveta and ultimately a series of predators of which the most numerous are fishing birds. Intermittently, this sequence is disrupted by the interpolation of a band of warm water along the coast, due to the "El Nino" effect, or Southern Ocean (ENSO) anomaly, causing a break in anchoveta production and creating havoc among the bird populations. Highly pulsed energy input maintains simplicity, and this is compounded by the rapid removal of the fish by the fishing birds. The intermittent destruction of the system prevents evolution of a more diverse ecosystem. All factors combine to maintain simplicity and high production between catastrophic events. There is a similarity between these events and agricultural activity, in that simplicity is maintained agriculturally by frequent disruption and rapid removal of the product.

In general high production in the oceans occurs in regions of upwelling. This is generally attributed to the bringing to the surface of nutrients from the deeper waters below the photic zone. Similarly, periods of high production are experienced in temperate lakes during fall and spring turnover. Again this is generally attributed to nutrient regeneration and transport. However, it has been shown by Legendre and Demers [224] through their work in the Gulf of St. Lawrence, that an increase in biological production may result from an input of physical energy, which they term "auxillary energy". Auxillary energy has the effect of a disturbing force, but without nutrient addition. It is apparent that an input of physical energy, breaking up the previously developed stability is sufficient to cause an increase in productivity. The established energy flow channels are disrupted and the system accumulates biomass energy in certain places in the food-chain. The energy previously utilized in the transfer of materials from one trophic level to another now appears in the form of increased biomass. This interesting aspect of the conversion of work done in the maintenance of stability into increased productivity, indicates the equivalence of the work done biologically maintaining stability and the physically work required to disrupt that stability. Again this process is fundamental to agriculture. Systems are simplified and production is increased by the work done in clearing, ploughing and subsequent cultivation. Agriculture is the imposition of severe boundary constraints maintaining top-down stability and homogeneity, but the moment the boundary conditions are relaxed, diversity increases as bottom-up stability increases.

## Conclusions

'I have yet to see any problem, however, complicated, which when you looked at the right way did not become still more complicated.'

(*Poul Anderson quoted by Koestler 1967* [148])

The merit of the theory of origin and structure of the biosystem outlined above is its essential simplicity. Fundamentally, there is a conflict between the tendency of energy to assume uniform distribution in accordance with the second law of thermodynamics and the asymmetry essential for the maintenance of energy flow. Close to thermodynamic equilibrium the two processes interfere with each other giving rise to a state of least dissipation. Most action (least dissipation) dominates the system behaviour in the short-term, least action dominates system behaviour in the long-term. Out of this asymmetry operating in different time frames, emanate the basic homeostatic mechanism and the long-term trends recognized as self-organization and evolution. "Evolution is opportunistic and immediate advantage is more potent than eventual gain" [225], but the opportunism is tempered by long-term constraints.

Ideally, a theory of the ecosystem based on thermodynamic principles, should be capable of generating testable hypotheses, but first and foremost it should be capable of explaining and condensing a large number of biological observations around simple general principles. A major factor in ecosystem structure is that biological temperature increases along the food-chain. A component of biological temperature is energy density, therefore, in agreement with Hirata and Fukao

[193], it was predicted that energy density should increase along the food chain. This was tested, and found to be true in a small arctic lake [178]. It is equally evident that the theory outlined brings together many concepts especially in the field of Production/Biomass ratio, the Biomass/Production ratio, Power/Economy ratio [179, 180], energy density, least dissipation and $r$- and $K$-selection under the umbrella concept of action.

Nevertheless, crossing interdisciplinary boundaries is a hazardous business, not lightly to be undertaken. Both sides are likely to shoot without challenge and recognition procedures. On the physical side, established principles must be conformed to and, if necessary, supplemented to meet the realities of biology, but additions must be built on the original precepts and must not replace them. This can be construed as nibbling sacred turf. On the biological side a strong philosophical element has to be accommodated and this is neither easy nor comfortable, but one does not do biology in a philosophical vacuum. The primary contribution of the theory to philosophy is resolution of the long-term conflict over teleology, the notion of progress toward an end represented by a 'perfect state.' The evidence that progress in the form of increased complexity and awareness of the world has occurred over evolutionary time has confounded and confused biologists since natural selection and the theory of evolution were first accepted. But as Simpson [226] states:

"Evolution is not invariably accompanied by progress, nor does it really seem to be characterized by progress as an essential feature. Progress has occurred within it but is not of its essence. . . . Within the framework of the evolutionary history of life there have been not one but many different sorts of progress."

The evolutionary trend to greater complexity is a function of the irreversible trend to greater energy flow, relative to biomass accumulated. Energy 'trapped' in the biosystem can 'escape' more rapidly only through greater complexity and greater diversity. When opportunity is presented through mutation the faster route is taken, and this emerges as progress. Thus teleology, as it expresses a specific 'end' (*telos*) is a misconception: the end is simply maximum power output.

The theory of ecosystem structure based on thermodynamic flows outlined above, demands a holistic approach to the biosystem. This approach is, in general, in conformity with much current thermodynamic thought [27], and with the concepts of systems analysis [10], and the philosophical views of Whitehead [227]. For this reason the study of ecosystems must start with simple autonomous systems.

Beyond the simplest level the biosystem rapidly assumes a state of near-infinite complexity. Unfortunately too, between the level of the whole system and the individual at the working level in an ecosystem, there is an enormous gap in which cause and effect become confused. The bridging of this gap is the greatest task facing biologists.

What may possibly be seen as the greatest stumbling block to acceptance of this thesis is the concept of the weak cohesive force. Yet cogent evidence for its existence is to be found in the coherent group behaviour of many varied animal taxa [228], and, although less frequently, in plants also [159, 229, 230]. It is possible to argue that this behaviour has evolved repeatedly on very many separate occasions, but the simplest explanation is that it is a magnification, whenever circumstances are appropriate, of an inherent potential, otherwise operating at relatively low levels. Without such cohesiveness it is extremely difficult to account for the relative stability of Earth's biosystem in the face of present perturbations to which it is being subjected. If density dependent mechanisms only had been operative the results would have been chaotic. Without self-damping the world hierarchy would not have rebuilt itself following displacement in times past. Philosophically, cohesiveness implies a constraint on independence and individuality, unacceptable as a concept to many.

Cohesiveness has allowed evolution in two directions. The first direction is toward a high degree of uniformity and concerted action, as displayed in schooling fishes such as the herring, *Clupea harengus*. Such close cohension demands inherent behaviour patterns, for there is little room for individuality. The second direction involves the formation of hierarchies within the general state of uniformity, such as is found in the wolf-pack. Coherence is retained through leadership and sub-ordination. In this way individualiy can be expressed without loss of overall unity. The formation of hierarchies within the group allows for a graduated response to external stimuli. Asymmetries, again, as at the world level, must not exceed certain values, for the "weak force" of cohesion is readily swamped by the "strong force" inducing least action.

An interesting expression of the weak cohesive force is found in highly diverse ecosystems. As diversity increased, the average interactional differential between species declined and, in certain cases where the interactional differential declined to near-zero, the cohesive force manifested itself in mutuality and synergism, such as displayed by the incredible adaptations of orchids to insect and bird fertilization, in the organized multi-species groups that develop in birds of the Amazon forest [231], and in the 'Mbuna fishes of Lake Tanganyika [53, 232, 233]. The system has evolved in a tightening spiral to the point where the extremes are in contact with each other: the initial symmetry of a single species of proto-organism ultimately gives rise to symmetry between species. The conclusion seems inescapable that, at base, a degree of coherence between all species must exist if a state of distorted symmetry is to be maintained.

The holistic point of view is greatly reinforced by consideration of the world's geochemical flows. As Reiners [115] emphasises: ". . . the world biota drives and regulates the global geochemical cycles of the major elements." A position developed by Lovelock [234] in the *Gaia* hypothesis. If the biological system originated in an anoxic environment, then change to an oxygen- based physiology would have been driven by the principle of least action, once an oxygen producing organism, had, by chance evolved. The world biota, in turn, is driven by the environmental signal system. The slight time asymmetry created a self-organizing homeorhetic mechanism.

The trends outlined above can perhaps, be most clearly identified by changes occurring in, and brought about by *Homo sapiens*. These changes have resulted from man's ability to modify his environment, with results only exceeded by the environmental modification brought about by the first oxygen producing organisms. The capacity for reasoning readily overcame the weak force, enabling the principle of least action to be harnessed to meet man's perceived needs. Virtually all material changes in man's living conditions have resulted from faster energy flows and more work done.

The assumption of dominance, initially gained through a large brain and manipulative hands, conferred on man the ability to utilize an ever-increasing range of "resources" which, in turn, allowed him to increase his dominance. "There is only one master generalist" [203]. Increased dominance stimulated increased energy flux at lower hierarchical levels, and the channelling of the increased production into human biomass. Communal wisdom emphasized the importance of increased energy flow and increased production at lower hierarchical levels. Dominance brought a greater increase in homogeneity, on the one hand through reduction of species diversity and the alienation of vast regions to the cultivation of relatively few species (even the cultivation of a few varieties of a single species); on the other hand, it brought about an immense increase in the human population.

Breaking down the barriers (both cultural and geographic) behind which specialized local communities evolved and survived, led to a much greater exchange of people and goods. This exchange, while of undoubted benefit at many levels, has also led to much greater "leakage," the net flow of energy between subsystems. In this case the net flow has been, not toward the tropical regions, but toward regions of relatively higher entropy production.

The greatest difficulty likely to be encountered in resolving the problems arising from our basic biology is the enormous gap between events at the world, or macro level, and those affecting individuals at micro level (Orians 1980). Too frequently events in the realm of the micro bring advantage to the individual but are detrimental to the whole. However, the feedback control mechanism operates only in the long-term so that the time asymmetry allows accumulation of an 'environmental deficit' that must be paid for later.

The global environmental problems with which we are faced require global solutions. They are of a magnitude that cannot be dealt with in any short-term or easy way, nor can we identify a specific goal toward which we should be heading. But one thing is certain, the path demands resolution of the trends to succession and abscession. For survival we must reinstate the ascendancy of the trend to most action. This necessitates choices. The dim perception of these constraints and recognition of the fact that choices are necessary for continuity was probably the origin of morality. Morality thus has a biological basis; it the conscious attempt to build a bridge between the microstate and the macrostate, between the immediate individual advantage and the long-term general good. We appear to have little choice but to develop and appropriate moral code, reinforced by a nested hierarchy of conventions, if we are to live within the laws of nature.

## Acknowledgements

I gratefully acknowledge the help and support of Peter Vanriel in numerous aspects of this work, particularly with respect to the interpretation of "action" in the biosystem as well as correction of the final manuscript. I am indebted to Gordon Koshinsky for the concept of the "environmental deficit." As always, I thank my wife, Cecile, for her patience and forbearance in the face of long field seasons and many hours of seclusion.

## References

1. Engleberg, J. and L.L. Boyarsky. 1979. The non-cybernetic nature of ecosystems. *Am. Nat.* 14:317–324
2. Darwin, C. 1959. *On the origin of species by natural selection.* First Edn. Reprinted 1968 (J.W. Burrow ed.), Penguin Books, London.
3. Gates, D.M. 1968. Toward understanding ecosystems. *Adv. Ecol. Res.* 5:1–36.
4. Watt, K.E.F. 1971. Dynamics of Populations: a synthesis. In P.J. Den Boer and G.R. Gradwell (eds.), *Dynamics of populations*, p. 568–580. Centre for Agricultural Publshing and documantation, Wageningen.
5. McIntosh, R.P. 1987. Pluralism in ecology. *Ann. Rev. Ecol. Syst.* 18: 321–341.
6. McIntosh, R.P. 1980. The relationship between succession and the recovery process in ecosystems. In J. Cairns, Jr. (ed.), *The recovery process in ecosystems* , pp. 11–62. Ann Arbor Science, Ann Arbor, Mich.
7. McIntosh, R.P. 1980. The background to some current problems of theoretical ecology. *Synthesis*, 43:195–255.
8. McIntosh, R.P. 1985. *The background of ecology: concept and theory.* Cambridge Univ. Press, London.
9. Hirata H, and R.E. Ulanowicz. 1986. Large-scale systems perspectives in ecological modelling and analysis. Ecol. Modelling 31: 79.
10. Patten, C.B. 1982. Environs: Relativistic elementary particles for ecology. *Am. Nat.* 119:179–219.
11. Patten, C.B. 1978. Systems approach to the concept of environment. *Ohio J. Sci.* 78: 206–222.
12. Patten, C.B., and E.P. Odum. 1981. The cybernetic nature of ecosystems, *Am. Nat.* 118: 886–895.
13. Patten, C.B, and G.T. Auble. 1981. System theory of the ecological niche. *Am. Nat.* 117: 893–921.
14. Forbes, S.A. 1887. The lake as microcosm. *Bull. Sci. Assn. Peoria, III.* 1887: 77–87.
15. Clements, F.E. 1916. *Plant succession: An analysis of the development of vegetation.* Carnegie Inst. Washington, Washington, Publ. No. 242.
16. Clements F.E. 1936. The nature and structure of the climax. *J. Ecol.*, 24:252–284.
17a. Tansley A.G. 1935. The use and abuse of vegetational concepts and terms. *Ecology*, 16:284–307.
17b. Tansley, A.G. 1923. *Practical plant ecology.* Dodd, Mead, New York.
18. Odum, E.P. 1969. The strategy of ecosystem development. *Science*, 164: 262–270.
19. Odum, E.P. 1971. *The fundamentals of ecology.* 3rd Edn. W.B. Saunders, Philadelphia.
20. Dunbar, M.J. 1972. The ecosystem as a unit of natural selection. In E.S. Deevey (ed., *Growth by Intussusception: Ecological essays in honour of G.E. Hutchinson*, pp. 113–130. Trans. Connecticut Acad. Sci., New Haven.
21. Coveney, P. V. 1988. The second law of thermodynamics: entropy, irreversibility and dynamics. *Nature* (Lond.) 333: 409–415.
22. Wicken, J.S. 1980. A thermodynamic theory of evolution. *J. theor. Biol* 87: 9–23.
23. Wicken, J.S. 1983. Entropy, information, and non-equilibrium evolution. *Syst. Zool.* 32: 438–443.
24. Wicken, J.S. 1985. Thermodynamics and a conceptual structure of evolutionary theory. *J. theor. Biol.* 117:363–383.
25. Wicken, J.S. 1984. Autocatalytic cycling and self-organization in the ecology of evolution. *Nature and System* 6:119–135.
26. Wicken, J.S. 1986. Entropy and evolution; ground rules for discourse. *Syst. Zool.* 35: 22–36.
27. Wicken, J.S. 1988. Thermodynamics, evolution and emergence: Ingredients for a new synthesis. In B.H. Weber, D.J. Depew, and J.D. Smith (eds.) *Entropy, evolution, and information: new perspectives on physical and biological evolution*, pp. 139–169. M.I.T. Press, Cambridge, Mass.

28. Mayr, E. 1976. *Evolution and the diversity of life*. Bellknap.
29. Oksanen, L. 1988. Ecosystem organization: Mutualism and cybernetics or plain Darwinian struggle for existence. *Am. Nat.* 131: 424–443.
30. Simpson, G.G. 1953. *The major features of evolution*. Columbia Univ. Press, New York.
31. Simpson, G.G. 1969. The first three billion years of community evolution. In: G.M. Woodwell, and H.H. Smith, (eds.), *Diversity and stability in ecological systems*, p. 162–177. Brookhaven Symp. Biol. 22.
32. Eldredge, N., and S. J. Gould. 1972. Punctuated equilibria: an alternative to phyletic gradualism. In: T.J.M. Schopf, (ed.), *Models in palaeobiology*, pp. 82–115. Freeman, San Francisco.
33. Stanley, S.M. 1981. *The new evolutionary timetable*. Basic Books, New York.
34. Gould, S.J. and N. Eldredge. 1977. Punctuated equilibria: the tempo and mode of evolution reconsidered. *Paleobiology*, 3: 115–151.
35. Boucot, A.J. 1983. Does evolution take place in an ecological vacuum? *J. Paleont.*, 57: 1–30.
36. Johnson, L. in press. The Far-from-equilibrium ecological hinterlands. In B.C. Patten, and S.E. Jorgensen (eds.), *Complex ecology*. Prentice-Hall, Englewood Cliffs, New Jersey.
37. Egerton, F. 1973. Changing concepts in the balance of nature. *Q. Rev. Biol.*, 48: 322–350.
38. Ross, J.C. 1847. A voyage of discovery and research in the southern arctic regions during the years 1839–1843. 2 Vols. John Murray, London. (Reprinted 1969, David and Charles, Newton Abbott, Devon).
39. Wallace, A.R. 1870. *Contributions to the theory of natural selection*, MacMillan, London.
40. Wallace A.R. 1878. *Tropical nature and other essays*. MacMillan, London.
41. Fischer, A.G. 1960. Latitudinal variations in organic diversity. *Evolution*. 14: 64–81.
42. Connell, J.H., and E. Orias. 1964. The ecological regulation of species diversity. *Am. Nat.*, 98: 399–414.
43. Pianka, E.R. 1966. Latitudinal gradients in species diversity: a review of concepts. *Am. Nat.*, 100: 33–46.
44. Sanders, H.L. 1969. Benthic marine diversity and the stability-time hypothesis. In G.M. Woodwell and H.H. Smith (eds.), *Diversity and stability in Ecological systems*. pp. 71–81. Brookhaven Symp. Biol. 22. Brookhaven, New York.
45. Slobodkin, L.B. and H.L. Sanders. 1969. On the contribution of environmental predictability to species diversity. In *Diversity and stability in ecological systems*, pp. 82–95. Brookhaven Symp. Biol. 22.
46. Menge, B.A. and J.P. Sutherland. 1976. Species diversity gradients: synthesis of the roles of predation competition and environmental heterogeneity. *Am. Nat.* 110: 351–369.
47. Schall, J.J. and E.R. Pianka. 1978. Geographic trends in number of species. *Science* 201: 679–686.
48. Mobius, K. 1877. Die Austern und die Austernwirtschaft. Verlag von Wiegandt, Hempel and Pary, Berlin. Trans. J.W. Rice, pp 683–751 in Report of the Commissioners for 1880. Part VIII, U.S. Commission of Fish and Fisheries.
49. Cowles, H.C. 1899. The ecological relations of the vegetation of the sand-dunes of Lake Michigan. *Bot. Gazette*, 27: 95–117, 167–202, 281–308, 361–391.
50. Clements, F.E. 1905. *Research methods in ecology*. Nebraska Univ. Publ. Co. Lincoln, Nebraska.
51. Whittaker, R.H. 1970. *Communities and ecosystems*. MacMillan, New York.
52. Southwood T.R.E. 1988. Tactics, strategies, and templets. *Oikos*. 52: 3–18.
53. Vannote, R.L., G.W. Minshall, K.W. Cumming, J.R. Sedell, and G.E. Cushing. 1980. The river Continuum. *Can. J. Fish. Aquat. Sci.* 37: 130–137.
54. Fryer, G., and T.D. Iles. 1972. *The Cichlid fishes of the Great Lakes of Africa*. Oliver and Boyd, Edinburgh.
55a. Lowe-McConnell, R.H. 1975. *Fish communities in tropical waters their distribution, ecology and evolution*. Longman, London.
55b. Lowe-McConnell, R.H. 1987. *Ecological studies in tropical fish communities*. Cambridge Univ. Press, Cambridge.
56. Simberloff, D. 1969. Experimental zoography of islands: A model for insular colonization. *Ecology* 50: 269–314.
57. Wilson, E.O. and D.S. Simberloff. 1969. Experimental zoogeography of islands: Defaunation and monitoring techniques. *Ecology*. 50: 167–278.

58a. Johnson, L. 1972. Keller Lake: characteristics of a culturally unstressed salmonid community. *J. Fish. Res. Board Can.*, 29:731–740.

58b. Johnson, L. 1976. The ecology of Arctic populations of lake trout, *Salvelinus namaycush*, lake whitefish, *Coregonus clupeaformis*, arctic char, *S. alpinus*, and associated species in unexploited lakes of the Canadian Northwest Territories. *J. Fish. Res. Board Can.*, 33:2459–2488.

59. Johnson, L. 1980. The Arctic charr. In E.K. Balon, (ed.), *Charrs, salmonid fishes of the genus Salvelinus*, pp. 15–98. Dr. W. Junk, The Hague.

60. Johnson, L. 1983. Homeostatic mechanisms in single species fish stocks in Arctic lakes. *Can. J. Fish. Aquat. Sci.*, 40: 987–1024.

61.' Platt, J.R. 1964. Strong inference. *Science*, 146: 347–353.

62. McLaren, I.A. 1964. Zooplankton in Lake Hazen and a nearby small pond, with special reference to *Cyclops scutifer*, Sars. *Can. J. Zool.* 42: 613–629.

63. Johnson, L. 1964. Marine-glacial relicts of the Canadian Arctic Islands. *Syst. Zool.*13:76–91.

64. Johnson, L. 1975. Distribution of fish species in Great Bear Lake with reference to zooplankton, benthic invertebrates and ecological conditions. *J. Fish. Res. Board Can.*, 32:1989–2005.

65. Holling, C.S. 1973. Resilience and stability in ecosystems. *Ann. Rev. Ecol. and Syst.*, 4: 1–23.

66. Johnson, L., and E.C. Gyselman (in prep.)

67. Ricker, W.E. 1975. Computation and interpretation of biological statistics of fish populations. *Fish. and Mar. Serv. Bull.* 191. Dep. Env. Canada.

68. Beverton, R.J.H., and S.J. Holt. 1957. *On the dynamics of exploited fish populations.* U.K. Min. of Agriculture and Fisheries Investigations (Ser. 2) 19, 533 p.

69. MacArthur, R.H. 1955. Fluctuations of animal populations and a measure of community stability. *Ecology*, 36: 553–536.

70. Elton, C.S. 1958. *The ecology of invasions by plants and animals*, Methuen, London.

71. Kolata, G.L. 1974. Theoretical ecology: beginnings of a predictive science. *Science*, 183: 400–401.

72. Elton, C.S. 1924. Periodic fluctuation in numbers of animals; their causes and effects. *Br. J. Exp. Biol.* 2: 119–163.

73. Elton, C.S. 1942. *Voles, mice and lemmings: problems in population dynamics.* Oxford Univ. Press, London.

74. Elton, C.S., and M. Nicholson. 1942. The ten-year cycle in numbers of lynx in Canada. *J. Animal Ecol* 11: 215–243.

75. Goodman, D. 1975. The theory of diversity-stability relationships in ecology. *Q. Rev. Biol.*, 50: 237–266.

76. Johnson, L. 1985. Hypothesis testing: Arctic charr, giant land tortoises, marine and freshwater molluscs and tawny owls. *Proc. Third Workshop in Arctic char.* I.S.A.C.F. Inst. Freshw. Res. Drottningholm.

77. Stoddart, D.R., and C.A. Wright. 1967. Ecology of Aldabra Atoll. *Nature* (Lond.) 213: 1174–1177.

78. Gaymer, R. 1968. The Indian Ocean giant tortoise, *Testudo gigantea*, on Aldabra. *J. Zool.* (Lond.), 154: 341–363.

79. Grubb, P. 1971. The growth, ecology and population structure of the giant tortoises on Aldabra. *Philos. Trans. R. Soc. Lond. B. Biol. Sci.*, 260: 327–372.

80. Bourn, D., and M. Coe. 1978. The size, structure and distribution of the giant tortoise population of Aldabra. *Philos. Trans. R. Soc. Lond. B. Biol. Sci.*, 282: 139–175.

81. Swingland, I.R. and M. Coe. 1979. The natural regulation of the giant tortoise populations on Aldabra Atoll. Reproduction. *J. Zool.* (Lond.), 186: 285–309.

82. Swingland, I.R., and C.M. Lessels. 1979. The natural regulation on Giant Tortoise populations of Aldabra Atoll. Movement, polymorphism, reproductive success and mortality. *J. Animal Ecol.* 48: 639–654.

83. Gibson, C.W.D., and J. Hamilton. 1983.Feeding ecology and seasonal movements of the giant tortoises on Aldabra Atoll. *Oecologia*, 56: 4–92.

84. Jones, E.W. 1945. the structure and reproduction of the virgin forest of the north temperate zone. *New Phytol.* 44: 130–148.

85. Southern, H.N. 1970a. Ecology at the crossroads. *J. Ecol.* 58: 1–11.

86. Southern H.N. 1970b. The natural control of a population of tawny owls, (*Strix aluco*). *J. Zool.*(Lond.), 162: 197–285.

87. Heuts, J.J. 1951. Ecology, variation and adaptation of the blind cave fish, *Caecobarbus geertsii*, Blgr. *Ann. Soc. R. Zool. Belg.*, 82: 155–230.

88. Poulson, T. L. 1963. Cave adaptation in Amblyopsid fishes. *Am. Midl. Nat.*, 70: 257–290.

89. Green, R.H. 1980. The role of the unionid calm population in the calcium budget of a small arctic lake. *Can. J. Fish. Aquat. Sci.*, 37: 219–224.

90. Hutchings, J.A., and R.L. Haedrich. 1984. Growth and population structure in two species of bivalves (Nuculanidae) from the deep sea. *Mar. Ecol. Progr. Ser.*, 17:, 135–142.

91. Sakai, A.K., and J.H. Sulak. 1985. Four decades of secondary succession in two lowland permanent plots in northern lower Michigan. *Am. Mid. Nat.*, 113: 146–157.

92. Hassell, M.P., J.H. Lawton, and R.M. May. 1976. Patterns of dynamical behaviour in single-species populations. *J. Animal. Ecol.*, 45: 471–486.

93. Thomas, W.R., M.J. Pomeranz, and M.E. Gilpin. 1980. Chaos, asymmetric growth and group selection for dynamical stability. *Ecology*, 61: 1312–1320.

94. Mueller, L.D., and F.J. Ayala. 1981. Dynamics of single species population growth: stability or chaos? *Ecology*, 62: 1148–1154.

95. Lindeman, R.L. 1942. The trophic-dynamic aspect of ecology. *Ecology* 23: 399–418.

96. Schrödinger, E. 1944. *What is life? The physical aspect of the living cell.* Cambridge Univ. Press, Cambridge.

97. Morowitz, H.J. *Energy flow in biology.* Academic Press, New York.

98. Johnson, L. 1981. The thermodynamic origin of ecosystems. *Can. J. Fish. Aquat. Sci.*, 38: 571–590.

99. Johnson, L. 1988. The thermodynamic origin of ecosystems. In B.H. Weber, D.J. Depew, and J.D. Smith (eds.), *Entropy, information and evolution.* pp. 75–105. MIT Press, Cambr. MA.

100. Wiley, E.O., and D.R. Brooks. 1982. Victims of history—a non-equilibrium approach to evolution. *Syst. Zool.* 31: 1–24.

101. Schneider, E.D. (1988) Thermodynamics, ecological succession, and natural selection. A common thread. In: Weber, B.H, Depew, D.J, and Smith J.D (eds) *Entropy, Information and Evolution*, M.I.T. Press, Cambridge, MA, p 107.

102. Prigogine I. 1978. Time, structure, and fluctuations. *Science*, 201: 775–785.

103. Nicolis, G. and I. Prigogine. 1977. *Self-organization in non-equilibrium systems: from dissipative structures to order through fluctuations.* Wiley Interscience, New York.

104. Prigogine, J., and I. Stengers. 1984. *Order out of chaos.* Bantam Books, New York.

105. Kleiber, M. 1961. *The fire of life.* Wiley, New York.

106. Lindstedt, S.L., and W.A. Calder, III. 1981. Body size, physiological time and longevity in homothermic animals. *Quart. Rev. Biol.*, 56: 1–16.

107. Calder, W.J. 1984. *Size, function and life history* Harvard Univ. Press, Harvard.

108. Paloheimo, J.E., and L.M. Dickie. 1966. Food and growth of fishes II. Effects of food and temperature on relation between metabolism and body weight. *J. Fish. Res. Board Can.* 23: 868–908.

109. Raup, D.M., and J.J. Sepkoski. 1982. Mass extinctions in the marine fossil record. *Science*, 215: 1501–1503.

110. Lewin, R. (1982) A downward slope to greater diversity *Science* 217: 1239–1240.

111. Newton, I. 1686. Rules of reasoning in philosophy. Reprinted in *Newton's Philosophy of nature*, H. Thayer (ed.), Hafner, New York.

112. Onsager, L. 1931. Reciprocal relations in irreversible processes. *Phys. Rev.*, 37: 405–426.

113. Soodak, H. and A. Iberall, A. 1978. Homeokinesis: a physical science for complex systems. *Science*, 201: 579–582.

114. Brooks, D.R., D.D. Cumming, and P.H. Leblond. 1988. Dollo's Law and the second law of thermodynamics: Analogy or extension? In B.H. Weber, D.J. Depew, and J.D. Smith (eds.) *Entropy, information and evolution*, pp. 189–224. M.I.T. Press, Cambridge, Mass.

115. Reiners, W.A. 1986. Complementary models for ecosystems. *Am. Nat.*, 127: 59–73.

116. Leigh, E.G. 1965. On the relationship between productivity, biomass, diversity and stability of a community. *Proc. Natl. Acad. Sci. U.S.A.*, 53: 777–783.

117. Whittaker, R.H., and G.M. Woodwell. 1972. Evolution of natural communities. In: J.A. Wiens (ed.) *Ecosystem structure and function.* pp. 137–159. Proc. 31st Ann, Biol. Colloquium, Oregon State Univ. Press, Corvallis.

118. Feynman, R.P. 1963. *The Feynman lectures on Physics,* (Vol. 1), R.P. Feynman, R. B. Leighton, and M. Sands (eds.), pp. 19–1–19–14. Addision-Wesley, Reading. Mass.

119. Davis, P. 1980. *Other worlds; Space, superspace, and the quantum universe.* Touchstone Books, Simon and Schuster, New York.

120. Watson, A. 1986. Physics — where the action is. *New Scientist,* January 1986, pp. 42–44.

121. Szent-Gyorgyi, A. 1961. Introductory remarks. In: W.D. McElroy, and B. Glass, (eds.), *Light is life,* p. 7–10. Johns Hopkins Press, Baltimore.

122. Goodwin, B.C. 1967. Biological control processes and time. *Ann. New York Acad. Sci.,* 138: 749–758.

123. Odum, H.T. and R.C. Pinkerton. 1955. Time's speed regulator the optimum efficiency for maximum power output in physical and biological systems. *Am. Sci.* 43: 331–343.

124. Waddington, C.H. 1968a. Towards a theoretical biology. In C.H. Waddington (ed.), *The basic ideas of biology.* Aldine Publ. Co., Chicago.

125. Polanyi, M. 1968. Life's irreducible structure. *Science,* 160: 1308–1312.

126. Anderson, P.W. 1972. More is different; Broken symmetry and the hierarchical structure of science. *Science* 177: 393–396.

127. Prigogine, I. 1973. Irreversibility as a symmetry-breaking process. *Nature* (Lond.), 246: 67–71.

128. Iberall, A., H. Soodak, and C. Arensburg. 1981. Homeokinetic physics of societies — a new discipline: autonomous groups, cultures, polities. In H. Real, D. Ghista, and G. Rau, (eds.), *Perspectives in biomechanics, Vol. i, Part I,* p. 433–527. Harwood Academic Press, New York.

129. Pierce, J.R. 1961. *Symbols, signals and noise: the nature and process of communication.* Harper and Row, New York.

130. Mayr, E. 1982. *The growth of biological thought.* Bellknap Press, Harvard.

131. Beer, S. 1972. *The brain of the firm: The managerial cybernetics of organization.* Allen Lane, Penguin Press, London.

132. Van Valen, L. 1973. A new evolutionary law. *Evol. Theory.* 1: 1–30.

133. McNaughton, S.J., and L.L. Wolf. 1970.Dominance and the niche in ecological systems. *Science,* 167: 131–139.

134. Stanley, S.M. 1973a. An explanation for Cope's rule. *Evolution,* 271: 1–26.

135. Saunders, P.T. and M.W. Ho. 1981. On the increase in complexity. On the increase in complexity in evolution, II. The relative complexity and the principle of minimum increase. *J. theor. Biol.* 90: 515–530.

136. Newman, M.J, and Rood, R.T. (1977) 'Implications of solar evolution for the Earth's early atmosphere,' *Science* 198: 1035–1037.

137. Stanley, S.M. 1973b. An ecological theory for the sudden origin of multicellular life in the late Precambrian. *Proc. Nat. Acad. Sci. USA,* 70: 1486–1489.

138. Stanley, S.M. 1974. Cropping and the Cambrian explosion. *New Sci.* 61: 131–133.

139. Golubic, S. 1976. Organisms that build stromatolites. In: M.R. Walter (eds.), *Stromatolites,* pp. 113–126. Elsevier, Amsterdam.

140. Jacobs, J. 1975. Diversity, stability and maturity in ecosystems influenced by human activities. In W.H. Van Dobben, and R.H. Lowe-McConnell (eds.), *Unifying concepts in ecology,* pp. 187–207. Dr. W. Junk, The Hague.

141. MacArthur, J.W. 1975. Environmental fluctuations and species diversity. In M.L. Cody, and J.M. Diamond (eds.), *The ecology and evolution of communities,* pp. 74–80. Bellknap Press, Harvard.

142. Southwood, T.R.E. 1977. Habitat, the templet for ecological strategies? *J. Anim. Ecol.,* 46: 337–365.

143. Dunbar, M.J. 1977. The evolution of polar ecosystems. In *Adaptations within antarctic systems.* Proc. third SCAR Symposium, Smithsonian Inst. pp. 1063–1076. Gulf Publishing, Houston, Texas.

144. Stehli, F.G., R.G. Douglas, and N.D. Newell. 1969. Generation and maintenance of gradients in taxonomic diversity. *Science* 164: 947–949.

145. Brown, J.H. 1973. Species diversity in seed-eating desert rodents in sand-dune habitats. *Ecology* 54: 775–787.

146. Abramsky, Z. 1988. The role of habitat and productivity in structuring desert rodent communities. *Oikos* 52: 107–114.
147. Connel, J.H. 1978. Diversity in tropical rain forests and coral reefs. *Science* 199: 1302–1310.
148. Koestler, A. 1967. *The ghost in the machine*. Hutchinson (Picador edn.), London.
149. Allen, T.F.H. and T.B. Starr. 1982. *Hierarchy perspectives for ecological complexity*. Univ. Chicago Press, Chicago.
150. Eldredge, N., and S.N. Salthe. 1984. Hierarchy and evolution. In *Oxford surveys of evolutionary biology*, R. Dawkins and M. Ridley (eds.), pp. 184–208. Oxford Univ. Press.
151. Salthe, S.N. 1985. *Evolving hierarchical systems*. Columbia Univ. Press.
152. O'Neill, R.V., D.L. De Angelis, J.B. Waide, and T.F.H. Allen. 1986. *A hierarchical concept of the ecosystem*, Princetown Univ. Press, Princetown.
153. Grene, M. 1987. Hierarchies in biology. *Am. Sci.* 75: 504–510.
154. Urban, D.L., R.V. O'Neill, and H.H. Shugart, Jr. 1987. Landscape Ecology. Bioscience 37: 119–127.
155. Waddington, C.H. 1968b. Towards a theoretical biology, *Nature* (Lond.) 218: 525–527.
156. Odum, E.P. 1977. The emergence of ecology as a new integrative discipline. *Science*, 195: 1289–1293.
157a. Gleason, H.A. 1926. The individualistic concept of the plant association. *Bull. Torrey Bot. Club.* 53: 7–26.
157b. Gleason H.A. 1939. The individualistic concept of the plant association. *Am. Mid. Nat.*, 21: 92–110.
158. Whittaker, R.H. 1953. A consideration of climax theory; climax as a population and pattern. *Ecol. Monogr.* 32: 41–78.
159. Harper, J.L. 1967. A Darwinian approach to plant ecology. *J. Ecol.*, 55: 247–270.
160. Mellinger, M.V. and S.J. McNaughton. 1975. Structure and function of vascular plant communities in central New York. *Ecol. Monogr.* 45: 161–182.
161. Whittaker, R.H. 1969. Evolution of diversity in plant communities. In G.M. Woodwell, and H.H. Smith, (eds.), *Diversity and stability in ecological systems*, pp. 178–196, Brookhaven Symp. Biol., 22, Brookhaven, New York.
162. Margalef, R. 1963. On some unifying principles in ecology. *Am. Nat.* 97: 357–374.
163. Margalef, R. 1967. Some concepts relative to the organization of plankton. *Oceanogr. Mar. Biol. Ann. Rev.* 5: 257–289.
164. Odum, E.P. 1960. Organic production and turnover in old field succession. *Ecology* 41: 34–49.
165. Golley, F.B. 1965. Structure and function of an old-field broomsedge community. *Ecol. Monogr.* 35: 113–137.
166. Loucks, O.L. 1970. Evolution of diversity, efficiency, and stability *Am. Zool.* 10: 17–25.
167. Reiners, W.A., I.A. Worley, and D.B. Lawrence. 1970. Plant diversity in a chronosequence at Glacier Bay, Alaska. *Ecology* 52:55–70.
168. Patrick, R. 1967. The effect of invasion rate, species pool and size of area on the structure of the diatom community. *Proc. Nat. Acad. Sci. U.S.* 58: 1335–1342.
169. Margalef, R. 1968. *Perspectives in ecological theory*. Univ. Chicago Press, Chicago.
170. Pickett, S.T.A. 1976. Succession: an evolutionary interpretation, *Am. Nat.* 110: 107–119.
171. Langford, A. and M.F. Buell. 1969. Integration, identity and stability in the plant association. *Adv. Ecol. Res.* 6: 83–135.
172. Frank, P.W. 1968. Life histories and community stability.*Ecology*, 49: 355–357.
173. Cooper, W.S. 1913. The climax forest of Isle Royale, Lake Superior and its development. *Bot. Gazette* 55: 1–44, 115–140, 189–235.
174. Watt, A.S. 1947. Pattern and process in the plant community. *J. Ecol.*, 35: 1–44.
175. Wright, H.E. Jr. 1974. Landscape development, forest fires, and wilderness management. *Science* 186: 487–495.
176a. Johnson, E.A. 1979. Fire recurrence in the subarctic and its implications for vegetation composition. *Can. J. Bot.* 57:1374–1379.
176b. Johnson, E.A. 1981. Vegetation organization and dynamics of lichen woodlands in the Northwest Territories. *Ecology* 62: 200–215.
177. Vogl, R.J. 1980. The ecological factors that produce perturbation-dependent ecosystems. In J. Cairns, (ed.), *The recovery process in damaged ecosystems*. Ann Arbor Science, Ann Arbor, Mich.

178. Vanriel, P. (1989). *Thermodynamic relationships between the trophic levels in a small autonomous lacustrine ecosystem in the Canadian Arctic.* MSc Thesis, University of Manitoba.
179. Gnaiger, E. 1983. Heat dissipation and energetic efficiency in animal anoxobiosis: economy contra power. *J. Exp. Zool.,* 228: 471–490.
180. Gnaiger, E. 1986. Optimum efficiencies of energy transformation in anoxic metabolism: the strategies of power and economy. In P. Calow (ed.), *Evolutionary physiological ecology,* pp. 7–36. Cambridge Univ. Press, London.
181. Elton C.S. 1927. *Animal ecology.* MacMillan, London.
182. Cousins, S.H. 1985a. Ecologists are building pyramids again. *New Scientist* 4 July, 1985, pp. 50–54.
183. Cousins, S.H. 1985b. The trophic continuum in marine ecosystems: structure and equations for a predictive model. In R.E. Ulanowicz and T. Platt (eds.) pp. 76–93. Can. Bull, Fish. Aquat. Sci. 213. Dept. Fisheries and Oceans, Ottawa.
184. Sheldon, R.W., A. Prakash, and W.H. Sutcliffe Jr. 1972. The size distribution of particles in the ocean. *J. Fish. Res. Board. Can.* 34: 2344–2353.
185. Kerr, S.R. 1974. Theory of size distribution in ecological communities., *J. Fish. Res. Board Can.* 31: 1859–1862.
186. Platt, T. and W. Silvert. 1981. Ecology, Physiology, allometry, and dimensionality. *J. theor. Biol.* 93: 855–860.
187. Dickie, L.M., S.R. Kerr, and P.R. Boudreau. 1987. Size-dependent processes underlying regularities in ecosystem structure. *Ecol. Monogr.* 57: 233–250.
188. Sprules, W.G. 1988. Effects of trophic interactions on the shape of pelagic size spectra. *Verh Internat. Verein. Limnol.* 23: 234–240.
189. Bonner, J.T. 1965. *Size and cycle: an essay on the structure of biology.* Princetown Univ. Press, Princetown.
190. Fenchel, T. 1974. Intrinsic rate of natural increase: the relationship with body size. *Oecologia* (Berl.) 14: 317.
191. Ulanowicz, R.E. 1972. Mass flow and energy flow in closed ecosystems. *J. theor. Biol.* 82: 223–245.
192. May, R.M. 1973. Mass and energy flow in closed ecosystems; a comment. *J. theor. Biol.* 39: 155–163.
193. Hirata H., and T. Fukao. 1977. A model of mass and energy flow in ecosystems. *Math. Biosci.,* 33: 321–334.
194. Hirata, H. 1980. A model of hierarchical ecosystems with migration. *Bull. Math. Biol.* 42: 119–130.
195. Brylinsky, M. 1980. Estimating the productivity of lakes and reservoirs. In E.D. LeCren and R.H. Lowe-McConnell (eds.), *The functioning of freshwater ecosystems,* pp. 411–453. Cambridge Univ. Press, Cambridge.
196. Saunders, G.W., K.W., Cummins, D.Z. Gak, E. Pieczyńska, V. Stråskrabová, and G. Wetzel. 1980. In E.D. LeCren and R.H. Lowe-McConnell (eds.) *The functioning of Freshwater Ecosystems,* pp. 341–392. Cambridge Univ. Press, Cambridge.
197. Winberg, G.G., V.A. Babitsky, S.I. Gavrilov, G.V. Gladky, I.S. Zakharenkov, R.Z. Kovalevskaya, T.M. Mikheeva, P.S. Nevyadomskaya, A.S. Ostapenya, P.G. Petrovich, J.S. Potaenko, and O.F. Yakushko. 1972. Biological productivity of different lake types. In *Productivity problems of fresh waters,* Z. Kajak, and A. Hillbricht-Ilkowska. Proc. IBP Symp. Kazimierz Dolny, Poland, May 6–12, 1970. Polish Scientific Publishers, Warszawa.
198. Ryszkowski, L. 1975. Energy and matter economy of ecosystems. In W.H. van Dobben and R.H. Lowe-McConnell (eds.), *Unifying concepts in ecology.* pp. 109–124. W. Junk, The Hague.
199. Mann, K.H., and M. Brylinsky. 1975. Estimating productivity of lakes and reservoirs. In T.W.M. 'Cameron and L.W. Billingsley (eds.), *Energy flow its biological dimension,* pp. 221–226. Royal Society Canada, Ottawa.
200. Golley, F.B. 1972. Energy flux in ecosystems. In: A.J. Wiens (ed.), *Ecosystem structure and function,* pp. 69–90. Proc. 31st Ann. Colloquium on Biol., Oregon State Univ. Press, Corvallis.
201. Whittaker, R.H. 1966. Forest dimensions and production in the Great Smokey Mountains. *Ecology* 47. 103–121.
202. Reiners, W.A. 1972. Structure and energetics of forests. *Ecol. Mongr.* 42: 71–94.
203. Burger, C. 1981. Why are there so many kinds of flowering plants? *Bioscience,* 31: 572–591.

204. Hartshorn, G.S. 1978. Tree falls and tropical forest dynamics. In: P.B. Tomlinson and M.H. Zimmerman (eds.) *Tropical trees as living systems*, pp. 617–638. Cambridge Univ. Press, New York.

205. MacArthur, R.H. 1972. *Geographical ecology*. Harper & Row, New York.

206. Sepkoski, J.J., R.K. Bambach, D.M. Raup, and J.W. Valentine. 1981. Phanerozoic marine diversity and the fossil record. *Nature* (Lond.), 293: 435–437.

207. Raup, D.M. 1986. Biological extinction in earth history, *Science*, 231: 1528–1533.

208. Alvarez, W., W. Alvarez, F. Asaro, and H.V. Michel. 1980. Extra-terrestrial cause for the Cretaceous-Tertiary extinction. *Science* 208: 1095–1108.

209. Alvarez, W., E.G. Kauffmann, F. Surlyk, L.W. Alvarez, F. Asaro, and H.V. Michel. (1984) Impact theory of mass extinctions and the invertebrate fossil record. *Science* 223: 1135–1141.

210. Jablonski, D. 1986. Background and mass extinctions: the alternation of evolutionary regimes. *Science*, 231: 129–133.

211. Northcote, T. G. 1988. Fish in the structure and function of freshwater ecosystems: a "top-down" view. *Can. J. Fish. Aquat. Sci.* 45: 361–379.

212. Hurd, L.E., M.V. Mellinger, L.L. Wolf, and S.J. McNaughton. 1971. Stability and diversity at three trophic levels in terrestrial successional ecosystems. *Science*, 173: 1134–1136.

213. Scott, W.B. and E.J. Crossman. 1973. *The fishes of Canada*. Fish. Res. Board Can. Bull. 184. Queen's Printer, Ottawa.

214. Holtzkamm, T., and M. McCarthy. 1988. Potential fishery for lake sturgeon, *Acipenser fulvescens*, as indicated by returns of Hudson's Bay Company Lac la Pluie District. *Can. J. Fish. Aquat. Sci.* 45: 921–923.

215. MacArthur, R.H., and E.O. Wilson. 1967. *The theory of island biogeography*. Princetown Monogr. Popul. Biol., Princetown Univ. Press, Princetown.

216. Pianka, E.R. 1970. On $r$-selection and $K$-selection. *Am. Nat.* 104: 592–597.

217. Parry, G.D. 1981. The meaning of $r$- and $K$-selection. *Oecologia* (Berl.) 48: 260–264.

218. Gomez-Pompa, A., C. Vazquez-Yanes, and S. Guavara. 1972. The tropical rain forest: a non-renewable resource. *Science* 177: 762–765.

219. Chevalier, R. 1948. Biogéographique et écologie de la fôret dense ombrophile de la Cote d'Ivoire. *Rev. Bot. Appl.* 28: 101–105.

220. Richards, P.W. 1952. *The tropical rain forest*. Cambridge Univ. Press, Cambridge.

221. Zaret, T.M., and R.T. Paine. 1973. Species introduction in a tropical lake. *Science* 182: 449–455.

222. Schaefer, M.B. 1970. Men, birds and anchovies in the Peru Current. *Trans. Am. Fish. Soc.* 99: 461–467.

223. Cushing, D.H. 1982. *Climate and fisheries*, Academic Press, London.

224. Legendre, L. and S. Demers. 1984. Toward dynamic biological oceanography and limnology. *Can. J. Fish. Aquat. Sci.* 41: 2–19.

225. Dobzhansky, T. 1968. On some fundamental concepts in Darwinian biology. In T. Dobzansky, M.K. Hecht, and W.C. Steere (eds.), *Evolutionary Biology*, pp. 1–34. Appleton–Crofts–Century, New York.

226. Simpson, G.G. 1949. *The meaning of evolution; a study of the history of life and its significance for man*. Yale Univ. Press, New Haven.

227. Whitehead, A.N. 1926. *Science and the modern world*. Cambridge Univ. Press, Cambridge (Reprinted 1985, Free Association Books, London).

228. McNaughton, S.J. 1984. Grazing lawns: animals in herds, plant form and co-evolution. *Am. Nat.*, 124: 863–886.

229. Janzen, D.H. 1976. Why bamboos wait so long to flower. *Annu. Rev. Ecol. Syst.*, 7: 347–391.

230. Werner, P.A., and H. Caswell. 1977. The population growth rates and age versus stage-distribution models for teasel (*Dipsacus sylvestris* Huds.). *Ecology*, 58: 1103–1111.

231. Munn, C.A. 1984. Birds of a different feather flock together. *Nat. Hist.* 93: 34–42.

232. Fryer, G. 1959. The trophic interrelationships and ecology of some littoral communities of Lake Nyasa, with especial reference to the fishes and a discussion of rock-frequenting Cichlidae. *Proc. Zool. Soc. Lond.*, 132: 153–181.

233. McKaye, K.R., and A. Marsh. 1983. Food switching by two specialized algae-scraping Cichild fishes in Lake Malawi, Africa. *Oecologia*, 56: 245–184.

234. Lovelock, J.E. 1979. *Gaia: a new look at life on earth*. Oxford. Univ. Press, Oxford.

# Environmental Systems

*G. H. Dury*

Sidney Sussex College. Cambridge, England

## Summary

Since the chapter title *Environmental Systems* is impossibly broad, this account confines itself to geomorphic systems, and within those largely to fluvial systems. Since the mid-century, fluvial morphology has been revolutionised in at least three ways: a fourth revolution may well be in progress, with a possible fifth in prospect. The launching of the analysis of stream network geometry in 1945 transformed work on network patterns. From 1953 onwards, discharge values were used in fluvial studies, effecting the hydraulic revolution and bringing magnitude–frequency relationships into prominence. Simultaneously, techniques of applied statistics developed largely in geology invaded geomorphology in force. One outcome of magnitude–frequency analysis is attention to events of great magnitude and low frequency, the relevant concept being that of neocatastrophism, which extends itself into palaeontology and into the study of natural hazards.

In the strict context of fluvial morphology, meanders remain problematic: no specific reason for meandering has yet been formulated. Against this, considerable progress has been made in calculating the discharges responsible for the palaeomeanders of c. 11–13 000 years ago. Longer-term considerations of climatic change apply to the former extent of deep chemical weathering, while arid environments supply evidence of frequently repeated changes of surface wetness.

Whether the strict systems idea, welcomed by some geomorphologists, will prevail generally is probably yet to be seen. Meanwhile, the possible fifth revolution, dependent on the application of modern higher mathematics, may have already begun.

## Introduction

Taken in its broadest sense, the above chapter title would permit very far-reaching treatment indeed—treatment that could include the whole of geomorphology and of ecology, much of geology, pedology, climatology, and meteorology, and sundry forms of applied study. In order to limit itself to what is manageable, the following text will confine itself to geomorphology, with particular reference to fluvial morphology. A further limitation is that emphasis will fall mainly on results produced in, or inspired by workers in, the United States of America. As will shortly be shown, geomorphology among Anglophones has been last revolution-ised by North American workers and their close associates. German studies can be shown to be very strongly influenced, as least in considerable part, by the concept of climatically-controlled morphogenesis of the landscape, a concept which, certain notable exceptions apart, has proved and remains somewhat suspect in the Anglophone world. French authors are known to have limited their readership by writing preferentially in their own language. While a widespread and indubitable lack of linguistic skill among Anglophones can only be regretted, it remains true that these have effected three, and possibly four, geomorphic revolutions during the last half century, and that a fifth may be in prospect. Let me hasten to add that, in expressing these opinions, I emphatically do not intend to denigrate the most valuable work done on the European mainland, the emergence of effective field studies in Africa, or the late but powerful arrival of Japanese scholars on the geomorphic scene. Results obtained in China, and to a great extent also those obtained in the U.S.S.R., are closed off from the rest of the world by linguistic barriers.

A definition of geomorphology as the study of the surface forms of the earth is not particularly helpful; but an extended definition that would be at all adequate would be impossibly long for present purposes. The scope of the subject, as perceived in the mid-1960s, is well illustrated by the 1240 pages of the Encyclopedia of Geomorphology [74]. Because of the overlap of geomorphology with kindred disciplines, the former can probably claim no subject matter of its own—with one noteworthy exception to be stated shortly. The diffuse character of geomorphology is demonstrated by the assortment of specialist disciplinary groups that are of interest to geomorphologists. For instance, the Geological Society of London has groups for British Geomorphological Research, British Sedimentological Research, Marine Studies, Tectonic Studies, and Volcanic Studies. Interest groups of the Geological Society of America include, among others, those concerned with Environmental Geology, Hydrogeology, Geomorphology–glacial geology, and Structural Geology-tectonics. Furthermore, individual geomorphologists may pursue far from rectilinear careers within a group of disciplines, or may switch lines altogether. Instances that come to mind are those who, beginning in geomorphology, have eventually settled in university and governmental administration, in cartography, in pedology, in environmental archeology, in environmental management, in vulcanology, in water quality studies, and in the study of inner-city housing.

The diffuse character of geomorphology is further illustrated by the fact that relevant findings appear in at least 50 accessible serials, to the amount of tens of thousands of papers a year. Although something of an overview may be obtained from the annual Binghamton Symposia volumes, Earth-Science Reviews, the Supplementbände of the Zeitschrift für Geomorphologie, sundry conference collections, and the admirable Geomorphological Abstracts, one can still find it difficult to keep track of what is actually going on—not least because fundamental ideas may well be developed elsewhere than in geomorphology—or to identify at once that very small proportion of published papers that is of prime importance. In the following text, most papers cited will be fairly to very recent: each citation should therefore be understood as implying *and references therein.*

The exception to the statement that geomorphology can claim no subject matter of its very own consists in the operation of earth surface processes in the medium term, say 25 to 250 years. The shorter term constitutes the domain of engineering time, the longer, that of geologic time. More on the time factor will be said later.

## Revolutions in Geomorphology

The development of geomorphology in the Anglophone world since the mid-century well exemplifies the comments of Kuhn [114] on scientific revolutions in general: 'A scientific revolution is a non-cumulative developmental episode in which an older paradigm is replaced in whole or in part by an incompatible new one. Normal science leads only to the recognition of anomalies [so that researchers make] strenuous efforts to force nature into the conceptual boxes supplied by professional education. [Since] any new interpretation of nature arises first in the minds of one or a few individuals . . . scientists are often intolerant of new theories invented by others.' A revolution is accomplished when its new wisdom becomes incorporated in textbooks.

The three certainly successful geomorphic revolutions of the mid-century related to stream networks, the use of data on stream discharge (the hydraulic revolution), and quantification. The possible fourth revolution relates to a combination of data-processing with the application of general systems theory. The potential fifth revolution, which like the fourth will be discussed later, depends on the application to geomorphology of modern higher mathematics.

The three certain revolutions have transformed geomorphology from the descriptive and historical study of landscape into the quantified study of process-response. The network revolution can be precisely dated to R.E. Horton's lengthy paper of 1945 [99]. As with the definition of geomorphology, so with Horton's paper: a one-sentence summary would not be useful. What Horton did was to show that network geometry—the number of streams in a given hierarchical order, the mean stream length per order, and the ratio of number between one order and the next—is essentially independent of geologic structure, obeying instead geometric laws. Two results immediately followed: a new kind of regularity in nature became apparent; and many fluvial morphologists, beginning with A.N. Strahler and his group at Columbia University, seized on the ordering system as the basis of comparison between channel and channel or basin and basin.

The speed with which the network revolution succeeded, especially after a slight modification of the ordering system by Strahler [157], is a guarantee of its timely instigation; an earlier and strictly relevant paper by Horton [98] appears to have made little impact. While network studies scarcely lend themselves to the generation of full-length textbooks, illustrations based on Hortonian principles began to appear in texts, including elementary texts, from about 1960, while an abridged form of the 1945 paper was included in a text of 1970 [58].

As Abrahams [1] has observed, many of the earlier network studies served to promote the idea of a drainage basin as an open system. A dramatic change occurred in 1966, when Shreve [151] introduced the ideas of links and of the random topology model. James and Krumbein [103] in 1960 distinguished between cis and trans links. Considerable further discussion followed up to about 1981, largely in the pages of *Water Resources Research*. A persistent difficulty has been that of defining network limits at the first-order level: that is to say, where can a stream, that does not issue from a spring, be said to start? Moreover, even in the short term, it can easily be imagined that drainage nets could undergo expansion and contraction similar to those known, on a larger scale and in a longer term, for networks subjected to increases or reductions in annual precipitation. But such problems aside, network analysis did turn geomorphological attention away from progressive adjustment to geologic structure, and toward internal regularities and considerations of possible randomness.

The planimetric patterns considered in network geometry are accompanied by profile patterns formed by sections along channels from sources to mouths. Consideration of stream profiles is entangled with that of strandline changes, which will be treated later. Ever since the French profile surveys of the 1800s, it has been clear that many profiles tend to be concave-upward. The reason, that channel slope tends to decrease as order increases, was for many years not understood, even in the alternative form that, as drainage area increases, so does channel-forming discharge, with a consequent reduction in slope. Also not understood was the fact that breaks in the profile, where slope increases for a time in the downstream direction, need not necessarily result from changes in ocean level (cyclic nickpoints) or from crustal movements (tectonic nickpoints), but can also reflect changes in channel efficiency. Whereas, in an ideal stream, efficiency would increase throughout in the downstream direction, mainly in response to the increase in channel size, local variations in efficiency, related for instance to the properties of the bedrock upon which a stream flows, can produce local steepenings of the profile—the main class of non-cyclic nickpoints. Such nickpoints were demonstrated by Wolman [181] for Brandywine Creek, Pennsylvania. For profiles in general, Langbein and Leopold [117] have shown that the usually observed profiles, nickpoints always excepted, are intermediate between the equal-work shape, where slope is proportional to $q^{-0.5}$, and the least-work shape, where slope is proportional to $q^{-1.0}$, q in each case being channel-forming discharge.

If a single paper can be taken as the start of the quantitative revolution in geomorphology, it is that of Strahler on the dynamic basis [156]. This was however by no means the first advocacy of measurement in geomorphology or closely-related disciplines. Morisawa [124] recalls that water discharge measurements were made in ancient Roman times, at least as early as 100 CE; measurements of the heights of Nile floods go back thousands of years further still. Scree angles have been studied since at least 1842, while slope surveys go back to at least 1875. Statistical analysis on a large scale was not however introduced until the mid-century. According to Burton [29], the quantitative revolution in geography as a whole had already succeeded by 1963. A major statement is that of Chorley [33], who wrote in 1961 but, on account of publishing delays, did not see his long paper in print until 1966. He makes clear the debts owed by quantifying geomorphologists to Strahler, and to W.C. Krumbein of Northwestern University.

At the textbook level, Anglophone geomorphologists may not yet have been ready for the two theoretical analyses of Scheidegger [146, 147]. Alternatively to a textbook, a new specialist journal might be accepted as a sign of revolutionary success: *Geographical Analysis,* which is entirely quantitative/mathematical, began publishing in 1969.

An important characteristic of statistical analysis is that it demands data—for geomorphic purposes, numerical data. It has greatly encouraged field measurement of slope angles, beach gradients, and channel dimensions. Channel measurements combined with measurements of stream discharge to sustain the hydraulic revolution that was effected by L.B. Leopold and his associates at the U.S. Geological Survey.

Until their work began to appear, few discharge measurements had been used by Anglophone geomorphologists, apart from those supplied by irrigation engineers in the Indian subcontinent. For workers in the U.K., indeed, very few discharge measurements were available. Leopold and his group, using the abundant streamflow records collected by the Survey, were able to relate changes in channel dimensions—principally width, depth, and cross-sectional area—to changes in discharge, both through time at a given cross section, and along the channel for a fixed discharge frequency. The relevant series of U.S.G.S. Professional Papers began in 1953 with an enquiry into this precise matter. The new text that marked the success of the hydraulic revolution was the 1964 production of Leopold, Wolman, and Miller [119]. The use of the expression *fluvial processes* in the title is significant. More than any other single work, this has led fluvial geomorphology among Anglophones to be process-oriented. Nansen [128] considered that it introduced into fluvial geomorphology in Britain a period of unbridled empiricism that still prevails today. What that author says about three fluvial texts of the 1980s could be restated in Kuhn's terms, by saying that they exemplify normal science.

Recalling Kuhn again, as saying that new truth does not dawn by conviction, but because its opponents die out, and using the information already given, along with some additions, we might speculate on the length of time needed for a geomorphic revolution. Among Anglophones, and possibly also among Francophones, geomorphology remained unchanged as to methods and aims from about

1900 to about 1950. The network revolution, although quite rapidly effected, needed perhaps the 25 years from 1945 to 1970 to achieve its full effect. The statistical revolution, although successful in about a decade, still had to contend with vigorous guerila bands of opponents; moreover, it is still having some of its implications worked out, because of the speed at which some geomorphic processes work—too slowly, that is, for data to be collected rapidly. Similar comments apply to the hydraulic revolution, successful in no more than 11 years, but still obstructed in many respects by lack of data. Wegener's hypothesis of continental drift, formulated in the 1920s, and bitterly opposed in the Northern hemisphere, had to await the paleomagnetic studies of the 1950s to provide additional (and refined) evidence, and then the development of plate tectonics from the 1960s onward to provide a set of mechanisms. Bretz's well known work on the Channelled Scablands of the U.S. Pacific Northwest, with its hypothesis, formulated in the 1920s, of a sudden outburst from an ice-dammed lake, also met savage opposition, allied to what we have seen Kuhn describe as attempts to force nature into existing conceptual boxes; final and overdue acceptance of Bretz's view came in 1965, that is, at the end of some 40 years. The hypothesis of climatic change as a cause of regional stream shrinkage, first propounded in the 1950s, appeared to have shed its original label of heterodoxy, and to have acquired that of orthodoxy, by about 1980. There is obviously a considerable spread in the record, from 11 to 40 or more years; however, despite what will now be said about the apparently incomplete and only partially successful systems revolution, it might be guessed that yet another geomorphic revolution is now due.

### Applications of Systems Theory

If systems theory is effecting a revolution in geomorphology, it has already taken some 35 years without having yet completely succeeded. The concept goes back eventually to von Bertalanffy in 1950 [172]. In an important, though not necessarily the first, discussion of the significance of systems theory in geomorphology, Chorley [32], comparing closed and open systems, concluded that certain advantages accrue from open-system thinking about landscape; among these are concentration on form-process relationships, recognition of multivariation, liberalisation of views on landscape evolution, liberalisation of views on aims and methods, and attention to whole landscape assemblages.

A much fuller and more complex approach is that of Chorley and Kennedy [34]. Since the work in question is a textbook of 1971, it could be argued, following Kuhn, that the systems revolution in geomorphology had succeeded in a space of 21 years. Subsequently, as will be observed, additional texts designed on systems lines have appeared; there have also been individual papers of considerable importance. On balance, however, it seems fair to suggest that the application of the systems concept in geomorphology, in the strict form, has come to be overshadowed by the attention paid to a single kind of system, namely, the process-response system, and also that the typology of geomorphic systems is not yet finally agreed.

Chorley and Kennedy define a system as a structured set of objects and/or attributes, these consisting of components or variables that exhibit discernible relationships with one another, and that operate as a complex whole according to some observed pattern. Although this summary definition is no more instructive than is a one-sentence definition of geomorphology, much clarification is achieved when geomorphic systems are classified. Thus, Chorley and Kennedy recognise a functional classification into closed systems, with no mass or energy crossing the boundaries; isolated systems, which can exchange energy but not mass with their surroundings; and open systems, which can exchange both mass and energy. Most natural systems are open. A second classification, depending on internal complexity, is structural: it distinguishes morphological, cascading, process–response, control, and self-maintaining systems, progressing thence through plants, animals, man, social systems, and human ecosystems. Geomorphology obviously deals with morphological, cascading, and process–response systems, except insofar as control systems, responding as they do to some intelligence, are relevant to applied geomorphology. Environmental studies in the broad sense deal with ecosystems.

The stream networks analysed by Horton and his successors constitute morphological systems. Cascading systems receive inputs and make outputs. A cascading system may be viewed as a black box, with no consideration given to internal structure; as a grey box, with interest centred on a limited number of subsystems; or as a white box, for which an attempt is made to identify as many storages and flows as possible, in order to perceive how the internal structure produces a given output in response to a given input. An example of a black-box cascading system is the karst terrain examined by Bassett and Ruhe [13], for which beeline distances from sink to riser, and therefore net slope and net velocity, are known. But since nothing is known of the actual sink to riser paths, net slope and net velocity have little meaning: only inputs and outputs are certain.

Process–response systems are regarded by Chorley and Kennedy as formed by the intersection of morphological and cascading systems, the links being provided by morphological states. Linkage commonly involves negative feedback. Emphasis falls on the identification of the relationship between a process and the forms resulting from it. A sandy beach, analysed in respect of slope, grain size, and wave impact is a process–response system. The history of geomorphic research shows that many process–response studies have begun with response, attempting to work back therefrom to process.

A somewhat different classification of geomorphic systems is offered by Strahler [158]. That author follows W.H. Terjung in organising information used in systems into an hierarchy of five levels: collected data; sets of morphological variables; flow systems of energy and matter; process–response systems; and systems regulated by cybernetic feedback. This last effect occurs in agricultural, biological, psychological, and social systems. The second to fourth levels of Strahler's hierarchy clearly correspond to the morphological, cascading, and process–response systems of Chorley and Kennedy. At the same time, Strahler, listing some 90 systems variables, provides an alternative set of symbols for circuit diagrams and flow-control mechanisms (Figs. 1, 2). Examples of the use of these symbols in diagrams for the global radiation balance, the soil-water balance, and a

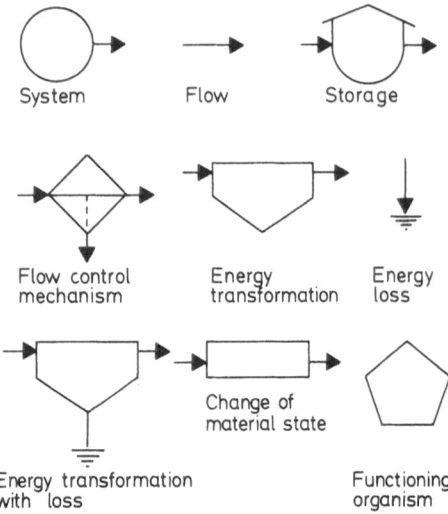

**Fig. 1.** Basis circuit symbols for diagramming energy and material flow systems

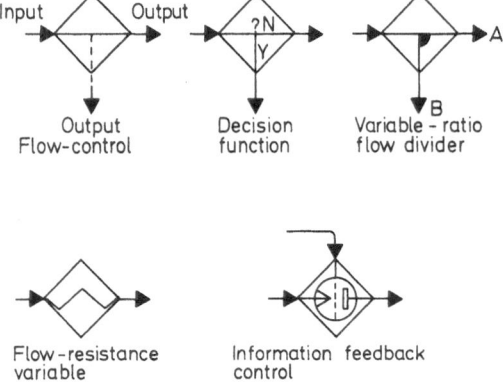

**Fig. 2.** Flow-control mechanisms used in circuit diagrams of energy or material flow systems

stream channel segment are given in Figs. 3–5. The systems variables in question are classified by Strahler into dynamic, mass-flow, geometric, and material-property.

The diagrams in Figs. 3–5 are system models. As Strahler remarks, systems exist in the real world, but models are artefacts: they provide the means of expressing and representing systems. Scale models, rational (deterministic), mathematical, empirical, analogue, and computer models are all employed by geomorphologists. Some noteworthy advances of recent years concern the application of computer modelling to the treatment of process–response models; although, as will be seen, this application is not without its dangers.

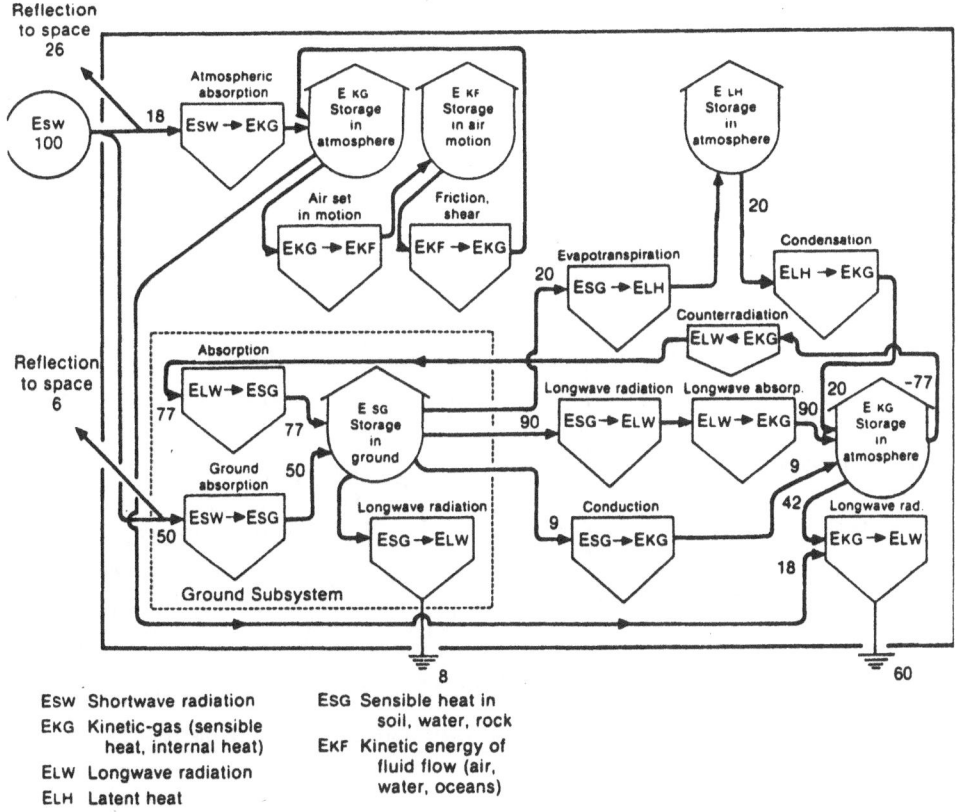

ESW   Shortwave radiation
EKG   Kinetic-gas (sensible heat, internal heat)
ELW   Longwave radiation
ELH   Latent heat

ESG   Sensible heat in soil, water, rock
EKF   Kinetic energy of fluid flow (air, water, oceans)

**Fig. 3.** The global radiation balance, an open energy flow system. Energy flow is stated in percentage units. Entering solar radiation at the top of the atmosphere is 100%, equal to 263 kilolangleys per year

Texts additional to those noticed, that incorporate the systems idea, are those of Dury [65], meant for beginners and thus possibly symptomatic of changed thinking, and the distinctly more rigorous work of Huggett [100], who concentrates on morphological, cascading (flow), and process–form (process–response) systems. Huggett also offers a second typology, of simple systems, complex but disorganised systems, and complex and organised systems. Examples, respectively, are the solar system, a hillslope mantle, and any geomorphic process–response system. He considers that conceptual models may lead to analysis by way of scale models, field sampling and statistical analysis, and mathematical modelling. He is especially instructive on statistical techniques and their application, deterministic models of glacier flow, mathematical models of slopes, and the application of catastrophe theory. Eventually coming to biotic models, he illustrates the general problem posed by the present chapter, of confining environmental discussion within reasonable bounds. He also observes that the dynamics so far employed in the relevant discussions are basically Newtonian, that relativistic dynamics are

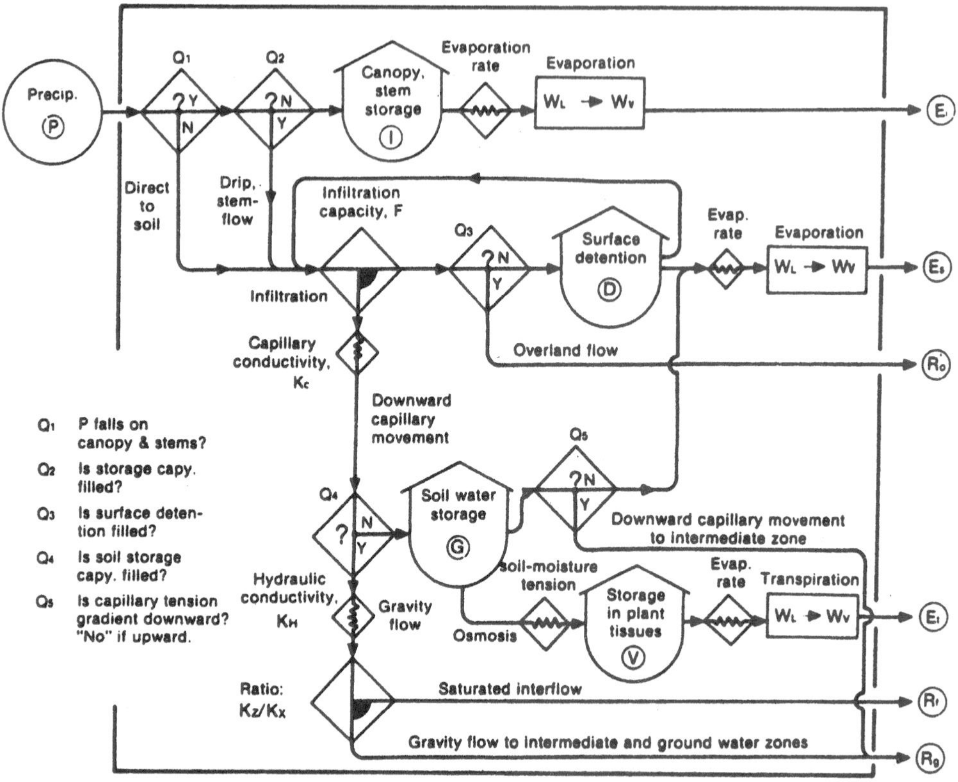

Precip. ($P$)

$Q_1$  $Q_2$

Evaporation rate — Evaporation — $W_L \rightarrow W_v$ — ($E_i$)

Canopy, stem storage ($I$)

Direct to soil

Drip, stem-flow

Infiltration capacity, $F$

$Q_3$  Surface detention ($D$)

Evap. rate — Evaporation — $W_L \rightarrow W_v$ — ($E_s$)

Infiltration

Overland flow — ($R_o$)

Capillary conductivity, $K_c$

Downward capillary movement

$Q_4$  Soil water storage ($G$)

$Q_5$

Downward capillary movement to intermediate zone

Hydraulic conductivity, $K_H$

Gravity flow

Osmosis

soil-moisture tension — Storage in plant tissues ($V$)

Evap. rate — Transpiration — $W_L \rightarrow W_v$ — ($E_i$)

Ratio: $K_z/K_x$

Saturated interflow — ($R_i$)

Gravity flow to intermediate and ground water zones — ($R_g$)

$Q_1$ P falls on canopy & stems?
$Q_2$ Is storage capy. filled?
$Q_3$ Is surface detention filled?
$Q_4$ Is soil storage capy. filled?
$Q_5$ Is capillary tension gradient downward? "No" if upward.

**Fig. 4.** The soil-water balance, an open material flow system

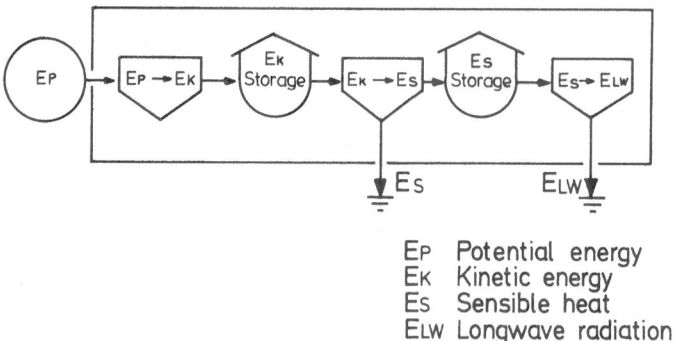

$E_P$

$E_P \rightarrow E_K$  |  $E_K$ Storage  |  $E_K \rightarrow E_S$  |  $E_S$ Storage  |  $E_S \rightarrow E_{LW}$

$E_S$   $E_{LW}$

$E_P$ Potential energy
$E_K$ Kinetic energy
$E_S$ Sensible heat
$E_{LW}$ Longwave radiation

**Fig. 5.** The stream channel segment as an open energy system

probably irrelevant, but also that quantum mechanics in the geomorphic context has not yet been explored.

Among individual papers in which the strict systems idea has been applied to geomorphology are those of Melton in 1958 [122] and of Brunsden [25] (Fig. 6).

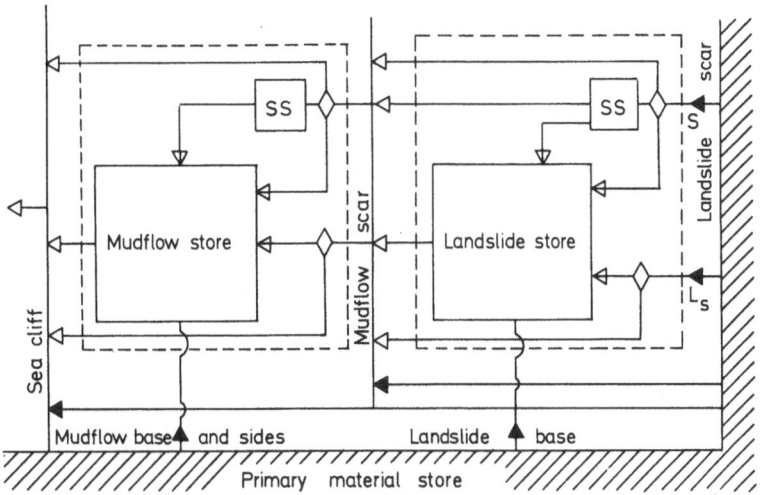

**Fig. 6.** The debris "cascade"

As this latter author observes, hillslopes may be considered as dynamic systems which have characteristic and measurable forms, materials with measurable properties, and energy relationships. Since external variables, particularly climatic variables, operate, it is possible to undertake statistical studies of relationships within systems, between systems, and between form and behaviour: a landslide system may be considered in terms of input, throughput, and output: by means of field measurements, slope budgets may be constructed.

The treatment of geomorphology in terms of systems runs parallel to the treatment of psychological perception in terms of gestalt: one works from the whole to the part, or at least to some of the parts. In this respect, systems treatment has distinct advantages over verbal treatment, since the written language is essentially linear.

Whether or not geomorphic matters are treated in strict systems terms, it seems likely that ever-increasing emphasis will be placed in future work on dynamic equilibrium/disequilibrium, negative/positive feedback, input/output relationships, storage, relaxation time, and thresholds of behaviour. Thresholds were the topic of the 1980 Binghamton Symposium [26], where however most of the contributions were site-specific. Among the general papers, that of Karcz [106] notes the rapid development over about a decade of the study of complex natural systems in physics, chemistry, biology, ecology, and topology, concluding that, despite their mathematical intricacy, the new ideas have much to offer to geomorphologists. The current steady-state approach contains limitations; transitions associated with increased entropy production lead to more complex states of organisation, which reflect the operation of self-organising tendencies.

Considerations of magnitude and frequency are already entangled with those of thresholds in the context of geomorphic step functions. Some systems diagrams are in effect computer flow charts (Figs. 7, 8). The spread of personal computers seems

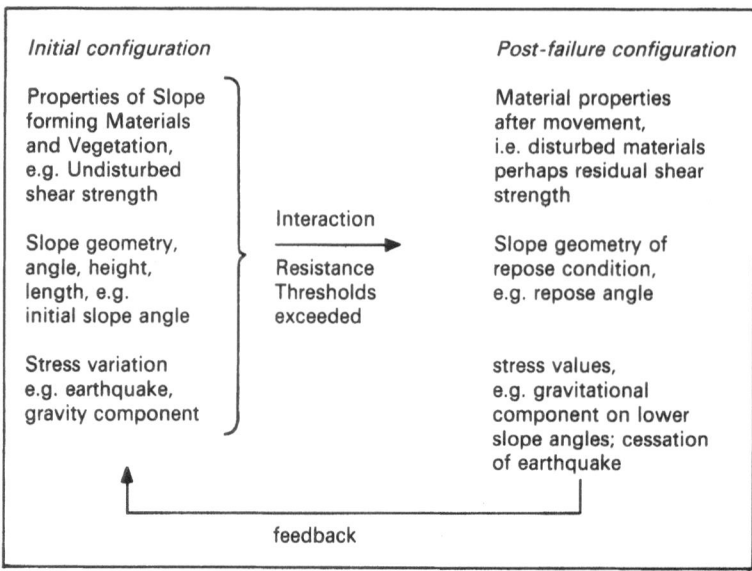

**Fig. 7.** Schematic diagram of a process–response model for mass movement

highly likely to lead to a great deal of empirical, hypothetical, and theoretical exploration of the ways in which geomorphic systems do, or conceivably can, work. Although Young [189] considers that systems theory has divorced much process-oriented work in geomorphology from process studies in other earth sciences, there can be little dispute that modern fluvial geomorphology is in actuality dominated by process–response studies.

**The Process–Response Approach in Various Settings**

Symptoms of the dominance of process–response are provided by a number of book titles and by at least one journal. The *Hillslope Form and Process* of 1972, by Carson and Kirkby [30], was rapidly followed by the *Drainage Basin Form and Process* of Gregory and Walling in 1973 [89]. In the space of nine years, that is, the main aspects of fluvial morphology had been treated in form-and-process texts. In 1976 the journal *Earth Surface Processes* was founded as the organ of the British Geomorphological Research Group, with an original declared scope of process measurement and mechanisms, process–response and simulation models, stability and equilibrium analysis, and magnitude–frequency relationships of events. The intended list of topics—sediment transfer, hydrology, weathering and surface geochemistry, soil dynamics and mechanics, and solute and nutrient cycling—is instructive as providing a possible summary of the perceived scope of geomorphic research. The primary intended readership included geologists, hydrologists,

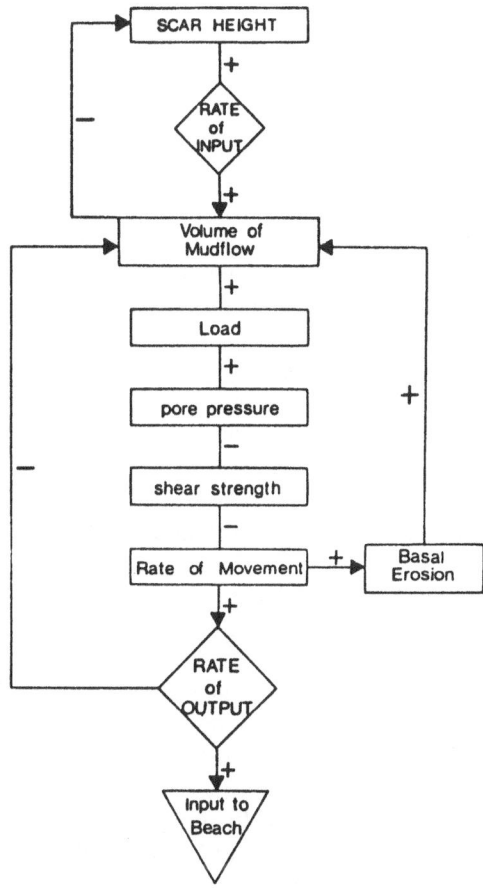

**Fig. 8.** A process–response system involving the form, stresses and rate of operation of a mudflow system

physical geographers, and sedimentologists, while agriculturalists, forest scientists, and pedologists were also thought likely to be interested to some extent. In order to provide for interests in landscape genesis and evolution, and also for applications to planning, environmental management, and engineering, the journal in 1981 changed its title to *Earth Surface Processes and Landforms* [111]. The process–form relationship was thus overtly recognised.

Also symptomatic of the dominance of process–response work is the 1981 collection edited by Goudie [82], in which the four parts deal respectively with form, the materials on which landforms occur, processes, and landform evolution and history. Processes are dealt with under the heads of weathering, slopes, solutes, river channels, glacial processes, and biological aspects of process measurement. In 1987 there appeared a collection on *Periglacial Processes and Landforms in Britain and Ireland* [18]; in the same year, the journals *Hydrological Processes* and *Geomorphology* began publication; the latter, founded as an outcome of the First

International Geomorphological Conference, may well do much to overcome the difficulties of cultural and linguistic barriers. Its scope is advertised to include the modelling of landforms, landform studies on all scales, extraterrestrial landforms, geomorphological processes, applied geomorphology, tectonic geomorphology, and climatological geomorphology. This last listing relates to the common sub-division of geomorphology according to territorial, lithological, or climatic con-text. A somewhat crude working distinction can be made among coastal, volcanic, karstic, glacial, periglacial, arid-zone, and fluvial geomorphology: such a distinc-tion cuts across, and is totally independent of, applications of systems theory, except that each subclass of geomorphology is assumed to be separated from the others by contrasts of process in addition to contrasts of form. Slope morphology stands somewhat apart from the rest; but it, too, has promoted vigorous discus-sions of process. More or less similar subdivisions occur in the collections edited by Pitty [138] and by Embleton and others [72]. Each subclass has at least one general text of its own.

Advances have by no means run in phase between one subclass and another; furthermore, great contrasts can be observed in the extent to which, under a given subhead, geomorphology draws on work in cognate disciplines.

Coastal morphology owes a very great deal to work at, and stimulus by, the Coastal Studies Research Institute at Baton Rouge, Louisiana, which as a centre of innovation ranks with those previously mentioned for fluvial morphology. This type of work lends itself to the application of energetics, which in the simplest form consists of the contrast between high-energy and low-energy beaches. To some extent it seems to have been diverted into historical tracks by the study of old beaches and related matters. The way in which coastal energetics may be analysed is well demonstrated by Tanner [162]. Identification of a cascading system in a coastal setting is in effect made by Cook [37], in a study of the known results of dune building where causes can only be postulated. Short and Wright [150] provide a process–response model for a high-energy beach, making use of a lengthy wave record to include dissipative, intermediate, and reflective beach types, according to modal wave height, wave period, and fall velocity of beach material (Fig. 9).

Considered in isolation, shoreline advance or retreat poses problems in the horizontal plane: but it cannot be separated from vertical displacements of the land, of the sea level, or of both combined. Relative displacement of sea level—strandline movements—have long been subdivided into positive, in which the level of the sea rises against the land, and negative, in which relative sea level falls. Postulated causes include epeirogenesis (continent-building), orogenesis (moun-tain-building), isostasy (differential loading/unloading), and eustasy (general movement, presumably worldwide). Subclasses of movement include the glacio-eustatic—referable to the waxing and waning of ice sheets—and the glacioisostatic, referable to crustal depression and rebound in response to glacial loading and unloading.

The rise of deglacial sea level to its present mark, about 5000 years ago, has partially or entirely submerged many former deltas, and has converted coastal inlets into estuaries, rias, and fiords. The results of negative strandline movements

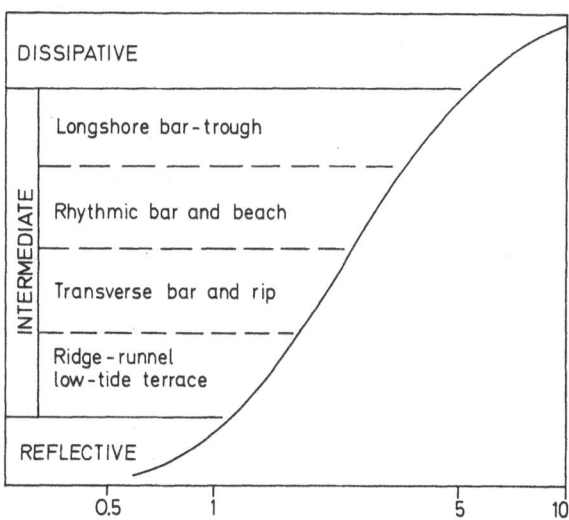

**Fig. 9.** Beach classification and subclassifiction in respect of wave action: adapted from Short and Wright [150]. The horizontal axis is scaled logarithmically in values of $(H_b/W_s)T$, where $H_b$ is modal wave height in m, $w^2$ is mean fall velocity in $ms^{-1}$, and T is wave period in seconds

are mainly accessible in certain orogenic areas, where their study is as yet in its beginnings, and in areas of deglacial isostatic rebound. The central part of the Scandinavian Peninsula, already uplifted through more than 200 m in the last 10 000 years, is still rising by as much as 1 cm/yr. Similar rebound in the north-eastern part of the North American continent has provoked complex changes in what is now the Great Lakes–St Lawrence drainage system. The most detailed set of evidence for deglacial rebound in North America is probably the array of emerged beach ridges on the eastern side of Hudson's Bay, for which Fairbridge and Hillaire–Marcel [76] find a nearly uniform rhythm of about 45 yr, extending back through 8 300 years and applying to a sequence of 185 ridges. These authors refer ridge spacing to the Double–Hale (*i.e.* four-sunspot period) cycle of solar magnetism, taking the coupling mechanism to be the stratigraphic generation of polar cirrus and low-latitude ozone, the two circumstances affecting the planetary albedo. It is not necessary to concur here to recognise the Fairbridge—Hillaire–Marcel model, or any other model of strandline change, as being designed in terms of process–response.

In Britain, the rise of Scotland and the north of England is accompanied by subsidence in the southeast: hence the need for the Thames Barrage. The typical geometry of deglacial rebound, combined with the sedimentary loading of the Rhine Delta, may not however be sufficient to explain the total observed effect: crustal dimpling, of which little is known, may also be in question.

In areas beyond the limit of deglacial isostatic rebound, a vast literature has accumulated on demonstrable or supposed eustatic movements—mainly negative movements, since ancient beaches and river terraces are for more easily accessible than are the sedimentary fills of drowned inlets. Difficulties arise when attempts are made to relate ancient beaches to river terraces, and these in turn to former strandlines. Still greater difficulties attend the identification of erosional platforms and the relation of these to former strandlines. This whole category of work will need to take into account the findings of Bryant [26] that, since 1960, the strandline has risen in some parts of the world while falling elsewhere. For the longer term, Vail and others describe numerous eustatic movements [171]. Such movements, now traced well down the stratigraphic column [105, 144, 149, 92], can have involved rises of sea level as great as 90 m/Ma during peak transgressions, and possible more rapid falls during peak regressions of as much as 170 m/Ma. These rates translate into 0.09 and 0.19mm/yr, surely enough to be geomorphically effective. Again, Ramsbottom in the paper cited recognises transgressive–regressive cycles of 1.1 to 1.6 Ma duration, with superimposed cycles averaging only 0.5 Ma—sufficient perhaps to affect some river profiles but not to disturb the large-scale processes of landscape planation. It is in the context of planation that the eustatic record impinges on gemorphology. A recent summary of the eustatic record for the last 255 MA by Haq and others [93] can be read as indicating an average of some 20 Ma for a transgressive–regressive sequence. On the shorter scale of Pleistocene time, Kukla [115] recognises at least 17 glacials and 17 interglacials in the last 1.7 Ma; these totals give a mean of 50 000 years for a glacial or an interglacial. On this basis, there has simply not been time for landscapes, or even streams, to adjust to glacioisostatic changes of ocean level.

In this light of findings such as these, it would appear that attempts to correlate erosional platforms over long distances will need to begin all over again.

As so often occurs in geomorphology, responses are defined before the responsible processes can be sought. Discussing causes of world-wide changes in ocean level, Donovan and Jones [48] conclude that variation in the total volume of mid-ocean ridges is too slow to produce the claimed effects, while rapid subsidence on continental margins is too low in magnitude. They look for an ultimate explanation in the history of mantle convection.

Scope for the study of volcanic landforms is spatially limited; in addition, the largest readily accessible extent of flood basalts, that in the U.S. Pacific Northwest, is interesting mainly as an area of cataclysmic flooding. Active volcanoes belong to the study of vulcanology, although lahars must be included among geomorphic events [e.g., 17]. The main relevance of karst morphology to geomorphology in general lies in the contributions its students have made to the radiometric dating of speleothems [89], with their implications for sea-level change, the dating of tills, and changes in climate. Not all karst terrain is developed on carbonates: for the East Kimberley region of northern Australia, Young [190] explains the diversity of karst terrain developed on arenites by antecedent leaching of silica combined with contrasts in rock strength and in modes of failure, these in turn being determined by porosities at the onset of weathering in the late Mesozoic or early Tertiary.

Insofar as it is concerned with glacial processes, glacial geomorphology is very largely dependent on the work of glaciologists—that is, on analysis in terms of physics. It is true that geomorphologists recognise the typical forms of glaciated country, and that they can claim the glacial breaching of highland divides as a special interest, on account of its impact on drainage patterns; but the whole topic of glaciation seems to belong to geologic rather than to geomorphic time. To cite only two of the available recent works, Barrett and others [12] extent the record of glacier ice on the Antarctic continent back to at least 29 Ma, while Imbrie [102] concludes that, on time scales from $10^4$ to $4 \times 10^5$ years, variations in the earth's orbit have been the main causes of fluctuations in the glacial record. On the shorter scale of $10^3$ to $3 \times 10^3$ years, the well established fluctuations of alpine glaciers relate to a cause yet to be identified. Within this range, however, geomorphologists have been able to contribute to knowledge with the aid of radiocarbon dating, lichenometry, and dendrochoronology. Some written historical records are also useful, as in the treatment of the Little Ice Age. As in other connections, so here: regional contrasts become evident. In place of the synchroneity of advances and retreats that might be expected, we have to deal with advances of glacier fronts in Norway from the late 1600s to the 1740s, and at the same time with glacier advances in the Swiss Alps mainly in 1600–10, 1690–1700, in the 1770s, and around 1820 and 1850, by which time the Norwegian glaciers had receded well away from their extreme Little Ice Age stands [90].

Possessing its own journal, *Biuletyn Periglaczialny*, periglacial morphology produces noteworthy studies of frost processes. To some extent, however, academic research seems to have been pre-empted by the engineering operations related to the Alaskan and Siberian pipelines. Perhaps as important as anything is the identification of relict periglacial features, well outside the existing limits of permafrost.

Arid-zone geomorphology makes slow and piecemeal advances. Of the 16 papers in a recent collection [131], the first four deal with wind-tunnel studies, while most of the others are site-specific. Neolson and Kocurek [130] refer the formation of star dunes to multidirectional primary windflow. Andersen and Hallett [5] develop a general model (clearly a process–response model) of aeolian transport: solutions for saltation and suspension are combined to yield mass-flux profiles. The combined solution shows saltation flux and suspension flux equal at about 0.8m above the surface, with suspension prevailing above this level and saltation below.

In some sense, arid-zone geomorphology is a subdivision of climatic geomorphology in general. The typical caution with which Anglophone geomorphologists appear to regard climatic geomorphology, that of glacial, periglacial, and arid regions always excepted, probably relates to the difficulties of obtaining data on the rates of operation of different processes in different climatic areas. Wang and Derbyshire [175], in a 575-km traverse from the Tibetan Plateau to the loess plateau of central China, involving a vertical range of about 3 000 m, recognise eight morphoclimatic zones—glacial, nivation, periglacial, semi-desert mountain gorge, desert mountain gorge, desert pediment–ephemeral river, desert sand dunes, and desert plain. The accompanying diagram (Fig. 10), constructed on the basis of their Table I, presents their conclusions about the relative importance of 12 kinds of process. These conclusions are based primarily on surface deposits and surface form. While no adverse criticism of their work is here intended, it is fair to say that the ideal—which must be very many years in the future—would be to determine known comparative rates of process. Rohdenburg [143], with reference to areas not subject to Recent or Pleistocene frost, recognises three types of planation process: retreat of high scarps, locally assisted by valley-floor pedimentation, in warm-arid regions; retreat of low steps, which were developed by incision in one climatic cycle, in seasonally-humid tropical regions; and dominant lateral erosion (equivalent to valley-floor pedimentation), subsequent to a climatically induced cycle of intensified incision, in humid-tropical regions. Deep weathering apart, however, most Anglophone obervers appear to be more impressed by the familiarity of inland tropical landforms than by their unfamiliarity.

With regard to hillside slopes, Cox [40], reviewing a symposium collection, notes the usual gradation between modellers without data and empiricists without theory. His strictures on modellers illustrate the danger, previously heralded, that can attend data processing in its application to some aspects of geomorphology. The empiricists he has in mind are presumably handicapped by the limitless variety that contrasts site with site, and also by the time necessary to arrive at any kind of generalisation.

In addition to Morisawa's observations, already cited, on early slope measurements, one also recalls that Charles Darwin emplaced markers to measure the

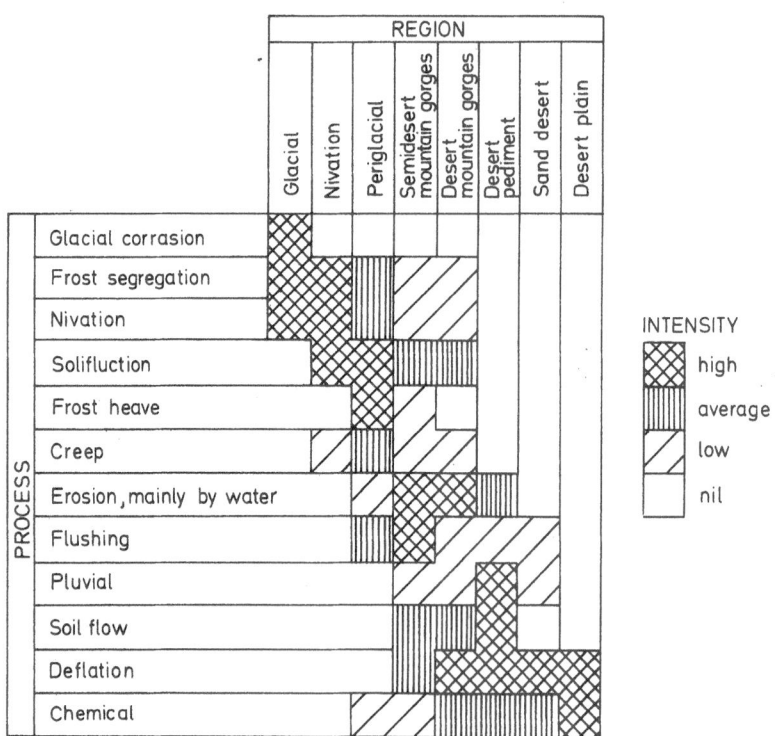

**Fig. 10.** Importance of processes in climatic geomorphology

speed of creep, unfortunately without recording site information for his successors. A great local geologist, Beeby Thompson, observed in the late 19th century that certain rock types are characterised by certain slope angles [167]. Among Anglophone geomorphologists, however, the dominant slope model until about the mid-20th century was that of convexo-concavity, with downwasting the primary control. It is difficult to reconcile such a model with, for example, the knife-edge erosional ridges developed under a humid climate in Tahiti. A.Wood [183] departed from the convexo–concave model in recognising a rectilinear segment, which he named the constant slope: but confusion has arisen as to whether Wood implied constancy in space—i.e., no more than rectilinearity—or constancy in time, i.e., backwearing with conservation of the slope angle. This whole matter is complicated by the vigorous advocacy of backwearing by L.C.King, and by mistaken attribution of the idea to Walther Penck, a subject well treated by Simons [152]. In practice, the whole debate about typical slope forms encouraged field measurements: those undertaken by Strahler and his group were related to positions in the stream (valley) network hierarchy.

When attention finally turned to processes acting on slopes, Rapp [141] made a field study of up to seven years of the action of rockfall, creep, and solifluction in Kärkevagge, North Sweden. Dalrymple and others [45] in 1968 propose a nine-unit slope model, wherein the form and development of each unit are controlled by a dominant process or set of processes (Fig. 11). While some of the ascriptions of process are scarcely to be doubted, others need testing, particularly with respect to their speed of operation. If a 25-year field study represents the minimum desirable, then it is clear that very little can yet have been done. Creep and solifluction are perhaps the easiest of slope processes to measure in operation; but even here, the

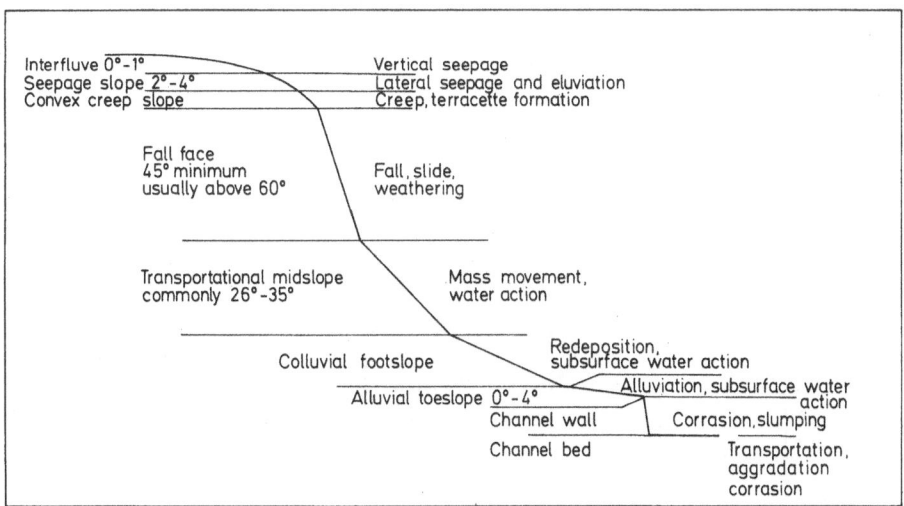

**Fig. 11.** The 9-unit slope model: units identified on the left, dominant processes on the right. Diagrams of this kind have appeared from time to time in sundry guises: the ultimate source is Dalrymple, Blong, and Conacher [45]

difficulties of arriving at generalisations are forbidding. Carson and Kirkby [30, 291–4], cite rates of creep that range from as little as 2 mm/yr at 0 cm depth and 0 mm/yr at 20–25 cm, to as great as 50–60 mm/yr at 0–2 m depth and 0 mm/yr at 10 m. Similarly, annual solifluction movement at the surface is known to range from 10 to 80 mm; future observation will presumably widen the range.

The study of rapid mass-movement, and especially of slumping, has been forestalled by students of soil mechanics, notably Terzaghi [165], on whose and generally similar work slope morphologists depend heavily.

Morphologists and pedologists have combined to identify on some slopes the operation of threshold-slope decline [31]. By this process, which relates to the crossing of a pedologic threshold, a rectilinear slope at a given angle is replaced from below by an equally rectilinear slope at a lesser angle (Fig. 12).

A special topic that has exercised numbers of slope morphologists is that of pediments. These are the concave-up features found at the bases of many slope systems. By historical accident, pediments have been studied mainly in areas of dry climate, where the processes at work on them have not yet been identified. But pediments are also found in areas of cool humid climate (Fig. 13), where the thickening of the soil mantle upslope of obstacles, and its thinning on the downslope side, show that the dominant process is creep. The theoretical implications of the fact that a pediment profile is usually a semilogarithmic curve have yet to be worked out.

More generally on the subject of process, Pilgrim and others [137] conclude from field studies that, contrary to expectations, the rates of slope development by processes of overland flow show no clear or simple relationship with mean annual rainfall, rainfall seasonality, vegetation type, lithology, or slope angle. Among site-specific factors that need to be taken into account are the frequency, duration, and intensity of rainfall, the height and proportional cover of vegetation, and a range of soil properties that includes infiltration characteristics; and even then, no clear pattern emerges. The authors suggest, on the basis of magnitude–frequency

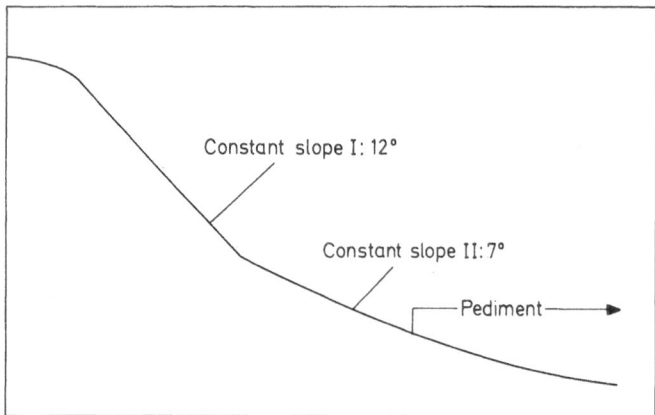

**Fig. 12.** Threshold–slope decline: vertical exaggeration × 5 based on a survey by the author

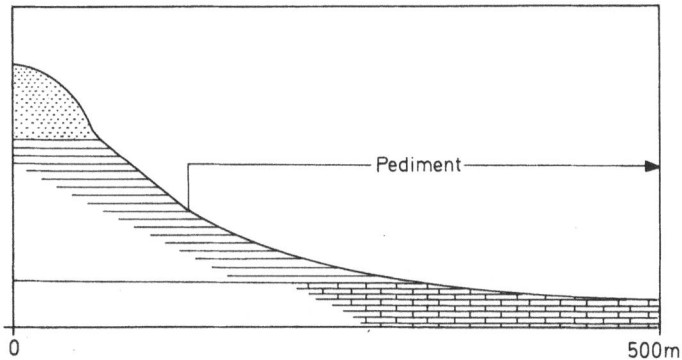

**Fig. 13.** A pediment profile developed across limestone and clay of part of the Jurassic succession in Midland England: vertical exaggeration × 5 based on survey by the author

considerations, that observations on hillslope plots need to extend through tens or even hundreds of years.

Rates of slope retreat are notoriously difficult to obtain: such few determinations as exist indicate a range of rate even wider than is known for creep or solifluction. In the longest-term study to date, Young and McDougall [186,191] use potassium–argon dates on Middle Eocene basalts to calculate the rate of retreat of steep valley walls in the Eastern Highlands of Australia; the rate lies in the range 10–25m/Ma, *i.e.,* 0.01–0.025 mm/yr. The coastal scarp, which is wholly erosional, has receded by a maximum rate of 170 m/Ma, but possibly much less, since the mid-Oligocene: *i.e.,* at a maximum of 0.17mm/yr, and this despite the influences of a rainforest climate and of rockfalls and other mass movements.

Prior to the mid-Oligocene, general uplift initiated a phase of canyon cutting. The extension of canyons by nickpoint retreat runs at a maximum of 5 km/Ma, *i.e.,* 5 mm/yr. Most rates of stream incision fall in the range 0.2–1.0 m/Ma, *i.e.,* 0.0002–0.001 mm/yr, with 0.008 mm/yr a maximum value [16].

A corollary of these findings on rates of retreat and incision is that some of the upland surfaces in the region were developed in the Mesozoic, and others in the Early Tertiary. For the Eocene surface, lowering by denudation is calculated at 0.66–1.8 m/Ma, *i.e.,* 0.00066–0.0018 mm/yr. Rates of operation of denudational processes, as inferred for the mainly soft rocks and glaciated and periglaciated areas of western Europe, may prove to be greatly exaggerated in the world view.

## Morphologic, Cascading, and Process–Response Systems in River Morphology

River morphology deals with stream channels: the more usual term fluvial morphology is often taken to include all the geomorphology of humid regions.

In respect of stream networks, it is unnecessary to add to what has been said earlier. In respect of channel morphology, a useful working distinction can be

made among straight, braided, meandering, anastomosing, and irregular channels
(Fig. 14). Straight channels, rare in nature except where firmly guided by lines of
faulting, are probably unstable. Braided channels display randomly arrayed mid-
stream bars of gravel, that shift and re-form in times of spate. The fundamental
cause of braiding appears to be a combination of coarse bedload and weak banks.
Although the slopes of braided streams are typically steep, by comparison with the
slopes of single-channel streams of equal discharge, the difference may be due to
nothing more than the approximate straight-line routes that braided streams take.
Anabranching channels, chiefly known, and chiefly, although as yet only initially
studied, on the Riverine Plains of inland southeastern Australia, are characterised
by offtakes. These latter, in some cases many kilometres long, either eventually
rejoin the parent channels, or unite with other main streams. Anabranching is
clearly favoured by very low cross-valley gradients. In this, it bears comparison
with some of the deltaic networks for which Morisawa [125] has analysed the
network geometry. Channels irregular in plan are simply those which at present
cannot be classified elsewhere. It is meandering channels that have attracted
principal attention.

If the T (key) and L patterns of the architectural fret decoration (Fig. 15) are
correctly identified by the alternative name of meander patterns, then abstract
morphological models of meanders date back to the IVth Dynasty of ancient
Egypt, some 4 500 years ago. Both of these patterns, and variants on them, were
widely used in classical Greece, about 3 000 to 2 500 years ago. Not until 1966 was
a satisfactory mathematical model for uniform meanders (the T type) established,

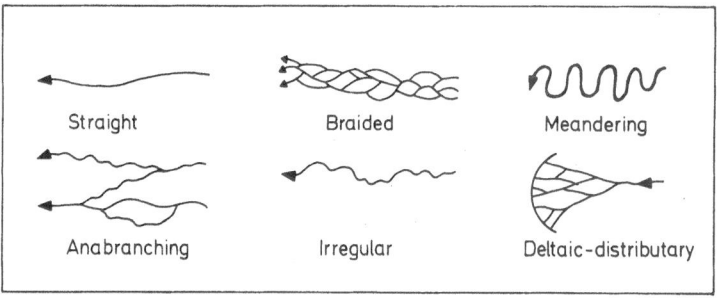

Fig. 14. Some common channel patterns

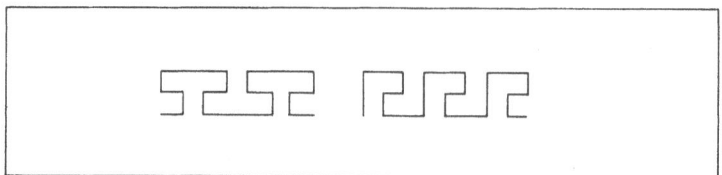

Fig. 15. The basic patterns, T and L, of the architectural fret, or meander

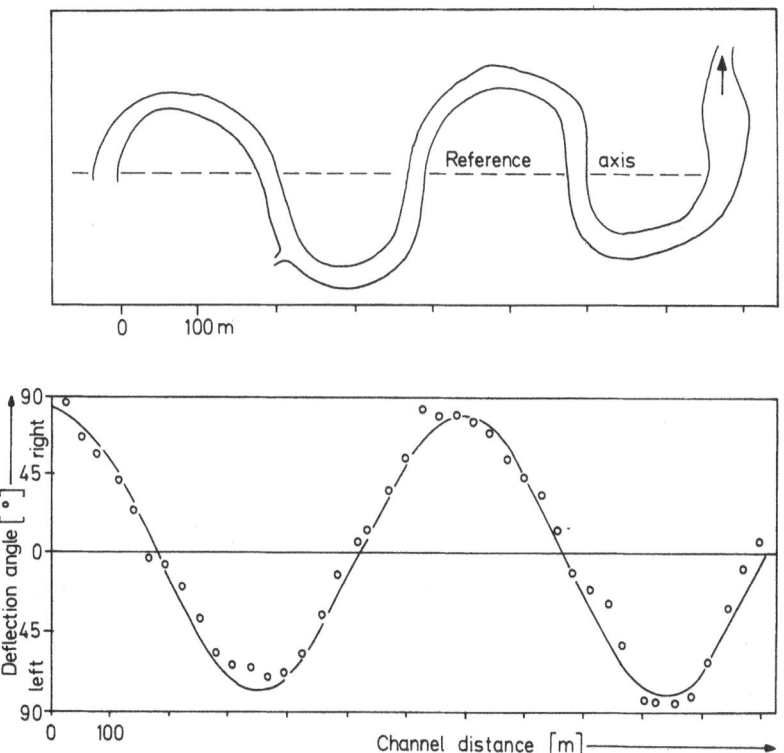

**Fig. 16.** The sine-generated curve: example from the Popo Agie River near Hudson, Wyoming, simplified from Langbein and Leopold [118]. Deflection angles are measured as departures from the reference axis. The equation for the curve on a single meander is $\phi = \omega \sin (s/M) \times 360°$, where $\phi$ is the channel direction at location s along the total meander channel distance M, and $\omega$ is the maximum deflection angle observed

in the form of a sine-generated curve [118] (Fig. 16). In the accompanying analysis, the authors conclude that depth, velocity, and slope are so adjusted as to decrease the variance of shear and the friction factor below the values that would obtain in a straight channel. It follows that meandering is the most probable form of channel geometry, and that it is more stable than straightness. As an alternative to analysis in terms of the sine-generated curve, Brice [22] advocates the resolution of irregular meanders into series of circular arcs, while Ferguson [78], dealing with asymmetric growth during migration (the L pattern), ascribes such growth to a spatial lag in kinematic relationships.

A vital question for river morphologists is why so many rivers meander at all. To this question there is as yet no clear answer. It is easy to show that, if a channel deviates from straightness, the meander habit minimises variation in width and slope (but not in depth), minimises the expenditure of work in flow round the bend, and also tends to minimise the deflection of the channel from the direct down-valley direction, in the sense that deflection angles taken along the channel tend strongly to be normally distributed round a mean and mode of zero. Some workers would conclude from this that the meander plan in essentially the product of random processes, drawing a comparison with the fact that stream network geometry can be simulated by random means.

The wavelength/width characteristics of meanders, such that wavelength commonly ranges between 8 to 12 bedwidths, has long been known. The explanation lies in the physics of flow round the bend: resistance is least when the radius of curvature (equivalent to half a wavelength) is between two and three bedwidths [8]. Analysis of a large array of data for midlatitude streams suggests that the central value for wavelength is about 11 bedwidths [62].

Meandering is by no means restricted to stream channels on the earth's lands. Meanders, cutoffs and all, are known for the Gulf Stream and for the midlatitude jet streams. They have also been detected on distributary channels on the Amazon Deep-sea Fan [79]. Meanders in the ocean and the atmosphere suggest that the numerous flume experiments made by engineers and geomorphologists may prove to be needlessly complicated, as incorporating a sediment load.

Little is known of the comparative frequency of the cutoff process, whereby meander loops are short-circuited, except that some meandering channels are highly liable to cutoff whereas others are not. To some extent, liability to cutoff may depend on irregularities of cohesiveness in floodplain alluvium: since such irregularities can impede the downstream sweep of meander bends, and can also cause individual loops to become distorted, they could be thought likely to increase the cutoff rate. In the range of geomorphic time, the shortening of the channel by cutoff seems likely to be balanced by the growth of new loops, so that channel length and channel slope are conserved. Thus, Laczay [116] finds for the Hernád River in Hungary that, without the artificial stabilisation of one potential cutoff, total channel length would have remained essentially stable in the interval 1957–72. Brice for the White River over 31 years [22], and Biedenhard and others for the Ouachita River over 80 to 160 years [15], testify to the long-term planimetric stability of some natural channels. Where meanders are articifially cut off, the result is a steepening of slope, usually followed by erosion of the bed.

Artificial cutoff on the lower Mississippi, downstream of Cairo, has from its beginning in 1929 reduced channel length by 12 per cent [95]. The concomitant and equal increase in channel slope has provoked the input of slugs of sediment, with ensuing bar formation and increased bank instability. Limiting navigation depths in the early 1980s were less than they had been 20 years earlier. Channelisation—straightening, often accompanied by enlargement of the cross section—on smaller streams has often proved disastrous. The data assembled by Keller and Brookes [110] show that channelisation can increase slope by a factor of up to 2.75, bankfull discharge by a factor of up to 11.13, and stream power—the factor here being a multiple of the factors for slope and discharge—by as much as 26.25 times. The inevitable result is erosion, tending toward a reduction of slope, on the channelised reach, downcutting on tributaries, and flooding and sedimentation further downstream.

Some clue to the control of the rate of meander migration is provided by Hickin in a study of the Squamish River in British Columbia [94]. He finds that the strength of secondary circulation (spiral flow) increases rapidly with increasing tightness of the bend. On tight bends, high shear stresses over the point bar and declining stresses at the concave bank markedly reduce the migration rate: at the extreme, separation zones developed at the concave bank may halt migration altogether or may even reverse it. With Nanson, the same author [96], using dendrochronology to measure migration rates for a 250-year period, concludes that these rates reach a maximum where the radius/width ratio is about 3.0: a prediction to the same effect could probably be made on the basis of the already cited work of Bagnold [8].

Hickin further observes that a river conveying a given discharge possesses the potential of at least eight significant degrees of freedom for change. Of these, width, depth, slope, and planform are morphological; the others are sediment calibre, sediment discharge, velocity, and boundary roughness. Measurements of width, depth, and velocity are complicated by the matter of dominant discharge. Information on channel slope is scanty, that on the discharge of bedload almost non-existent. Sediment transport equations are based mainly on flume studies: universal equations of this type may never become available. Since stability equations are open to the criticism of being black-box solutions, meandering systems must perforce be regarded as cascading systems: the most successful process-response work so far performed in this connection is probably that concerning short-term adjustments of bed and banks.

Difficulties in the treatment of bedload include the discontinuities that can occur at confluences [161], the fact that much bed material, including the coarse fraction, is often wash load for part of each high-energy event [163], and the further fact that bedload discharge can be influenced by rhythmic pulses that are not matched in the fluctuations of hydraulic variables [142]. It can be said, however, that bedload transport appears to cease when the mean value of stream power on the falling flood limb declines to 20 per cent of the value at which sediment motion is initiated. The complexity of process-response likely to be involved in the construction of riffles is shown by Winkley's indentification [179] of 18 factors as contributing to the control of riffle depth below the surface. Presumably an equal complexity applies to the construction of bars in a braided channel.

## Dominant Discharge and Magnitude–Frequency Analysis

Dominant, or channel-forming, discharge has been widely discussed in relation to meandering channels. If it could be precisely defined, it would provide an apt starting point for process–response enquiry. Such discharge is taken by many to be discharge at natural bankfull—that is, the discharge that would fill to the brim a channel that is neither progressively cutting not progressively filling. If the discharge could be related to channel dimensions, then it could be related to channel dimensions, then it could be reconstructed for paleochannels.

Bankfull discharge is clearly of modest magnitude and high frequency [61, 182]. The commonest form of magnitude-frequency analysis is that due to

E. J. Gumbel, which uses annual peak discharges to calculate recurrence interval as

$$r.i = (n + 1)/r$$

where r.i. is the recurrence interval of the event, in years, r is the rank order of a given event by magnitude, from the greatest downward, and n is the number of items in the series. The constant $+ 1$ merely ensures that the median event is placed at the half-way mark of the series. A more elaborate form of analysis, that of partial-duration, uses all events above a selected cutoff. It is a statistical property of Gumbel analysis that the mean annual flood, $q_{2.33}$, has a recurrence interval of 2.33 years, while the most probable annual flood, $q_{1.58}$, has one of 1.58 years. This latter value is equivalent to the 1-yr flood, $q_{1.0}$, on the partial-duration series. On this basis, and on the results of analysis of the recurrence intervals of known bankfull discharges, it has been argued that $q_{1.58}$ (annual series) can be taken as natural bankfull discharge, and that it provides a valid basis of comparison between reach and reach and river and river [59].

These statements should be read as applying to alluvial channels. Different considerations apply to flow in narrow gorges. For example, Baker and Pickup [11], analysing the flood geomorphology of the Katherine Gorge in the Northern Territory of Australia, find that the observed pool-and-riffle sequence is shaped by discharges far in excess of the most probable annual value; they cite the 100-year flood as an instance, with its discharge of about 6 000 m³/s, mean velocity through the cross section of about 7.5 m/s, stream power as great as $10^4$ watts/m², and bed shear sress of $1.5 \times 10^3$ N/m².

Williams [177], in a most useful treatment of the whole matter of bankfull discharge, obtains the empirical equation

$$q_{bf} = 4.4\ A_c^{1.20}\ S^{0.29}$$

where $q_{bf}$ is bankfull discharge in cumecs, $A_c$ is channel cross-sectional area in m², and $S$ is channel slope in the usual dimensionless form. While this equation accounts for 96 per cent of the sums of squares of log $q_{bf}$, it is still subject to a standard error of 42 per cent. In another test on known data [69], it has been shown to predict from 67 to 142% of observed discharge values at the 50% confidence level, and between 30 and 331% at the 95% level. Closely similar results are obtained by the use of other, although less simple, equations that use area or slope/area values. A frequent practical problem is that data on annual flood peaks, and thus the results of magnitude–frequency analysis, are far more readily available than are data on channel width, depth, and slope.

Magnitude–frequency analysis of annual peak discharges may not, in its usual form, be the most appropriate. Many long records of streamflow can be resolved into square waves, wherein two or three regimes—high and low, or high, medium, and low—can be distinguished [66]. This fact has yet to be explored in relation to dominant discharge. On the River Severn near Shrewsbury, England, bankfull flow through pools appears to be sensibly equal to $q_{1.58}$ on the total regime, whereas bankfull flow over riffles is equally close to $q_{1.58}$ on the low-peak regime [68]. Not only, therefore, may it be necessary to treat pools and riffles differently in respect of their magnitude–frequency relationships: enquiry is urgently needed into the implications of step-functional changes of regime for channel dimensions generally.

In this connection, some channels seem likely to prove better subjects than others—specifically, those which are highly sensitive to changes in discharge. At the other extreme come channels such as those of the Nene and Great Ouse in eastern England, which in the spring of 1947 experienced the flood event of recurrence interval in the range 500 to 1 000 years; but reshaping of the channels was confined to scour under bridge arches, the scoured material being dumped immediately downstream. Insensitivity is also displayed by the small streams investigated by Nanson and Young [129]; flowing across an alluvial plain slightly above sea level, these streams have failed to migrate laterally, while at the same time depositing fine material on their valley floors. As a channel rises, gravel bedload passes below the level of displacement, so that each channel is underlain by a wall-like accumulation of gravel that rises from the base of the fill to the bottom of the existing channel.

A high degree of sensitivity is by contrast exhibited by arroyos, the mainly dry incised washes of the American Southwest and elsewhere. These represent process–response systems where the response is obvious but the process is by no means agreed. Graf [86] lists 34 studies that emphasize grazing as an initiative cause, 20 that stress catastrophic floods, and 43 that appeal to climatic causes, including both increased wetness and increased dryness. Allowing for a certain overlap from list to list, we are still left with well over 50 studies but, as Graf says, without a widely accepted and well integrated theory of arroyo development.

In a further discussion, of erosion problems on the Navajo Indian Reservation, which in common with much of the U.S. Southwest experienced accelerated erosion from about 1880 onwards, the same author records that, to begin with, most authors thought overgrazing to be responsible [87]. By the mid-1940s, however, opinion had swung in favour of climatic change. For the test period 1930–60, Graf finds that stocking levels explain only 1 to 5 per cent of the variation in yields of water and sediment, whereas variations in surface moisture explain 38 to 66 per cent. Thus, although on general grounds one might expect the furtherance of investigation to transform process–response systems into control systems, where the environmental impacts of human activities are progressively better defined, Graf's findings for the Reservation seem to relegate an apparent, even if inadvertent, control system to the status of process–response.

In that the effect (response) is known, but the cause (process) in many cases is not, the topic of waterfalls resembles that of arroyos. The most effective single analysis is that of Philbrick [136] for the caprock model of Niagara Falls, wherein the definition of stress lines illuminates the phenomenon of rockburst. This however is not the only model possible; many actively retreating falls in Australia are buttressed outward at the base [188]. A closely related question is that of the abrupt change of channel habit, from wide and shallow, often with low rocky islets, upstream of the fall, to deep and narrow from the base of the fall downstream. Such change can persist through a long history of fall recession, as is shown among others by Niagara itself, the Grand Falls of the Potomac, and Victoria Falls. It can also be seen on a small scale, as at Pistyll Rhaiadr on the North Welsh border.

**Channels and Divides**

Although processes operating in channels are dominated by events of modest magnitude and high frequency, the same may not be true of processes operating on divides. In addition, except in such areas as the gullied loessland of the middle Hoangho basin, where sediment delivery at the mouth of a catchment can reach 100% of the amount mobilised within the catchment, it is usual to find delivery running lower than mobilisation. Walling and Webb [174] observe an exponential decline of delivery ratio with increasing size of basin, such that 10 to 50% of mobilised sediment is delivered at 100 km², but as little as 10% at 1 000 km². Slope deposition appears to be the major sediment sink, accounting for 30 to 60% of total disappearance: others are floodplains and channels. Lakes, where present, constitute automatic sediment traps [133]. Most of the studies of sediment production on divides relate to the short term. A longer view is provided by Graf [88] who, for the canyons of the Colorado Plateau, identifies two episodes of sediment storage, about 1250–1880 and from 1943 onwards. Since processes of sediment transport and storage are discontinous both in time and in space, generalisation, as in so many other fluvial connections, remains difficult; but the long intervals that can occur between mobilisation and (presumable) eventual delivery appear sufficient, in themselves, to suggest that the magnitude–frequency relationships applicable to channels may not be applicable to divides.

Processes acting on steep divides include landslips of the slump type, the usual trigger being heavy rainfall. The few studies so far available suggest that falls of 100- to 150-year recurrence interval are typically responsible. Lahars and other mudflows can involve movements that, in terms of the mass–distance index, are comparable with some annual totals of denudation for whole continents [61]. On the input side, single erupting volcanoes can eject as much material in a single year as is stripped annually off any continent except Asia. Once again, the magnitude–frequency relationships of channels, which emphasise modest magnitudes and high frequencies, may not be transferable to events that affect divides.

**Paleomeanders**

Very many existing streams, whether themselves meandering or not, are contained in meandering valleys. The term *underfit*, coined originally to connote meandering alluvial streams within the far larger curves of winding valleys, has been superseded by *manifest underfit*. The inference that underfit streams have experienced a reduction of discharge is sustained by field evidence, but the hypothesis that that reduction was due to stream capture is, in general, not. Since manifest underfits are widely, although not universally, distributed throughout the midlatitudes and beyond, their present condition must be ascribed to climatic change [50, 52, 53, 54]. Where capture of some other form of diversion has occurred, its effects can be disentangled from those of general shrinkage [51].

Since the former large meanders have been shown in numerous cases to be associated with former large channels, it becomes possible to use the former channel dimensions in attempts to reconstruct paleodischarges [62]; but the uncertainties of defining former banktops, added to the previously mentioned uncertainties of working back from dimensions to discharge, ensure that nothing better than broad approximations can be obtained: this holds, even when predictive equations are properly applied [178]. Nevertheless, it seems likely that the former large streams could have been nourished by mean annual precipitation some 1.5 to 2.0 times that of the present day. The most precise determinations of former discharge are those of Rotnicki [145], who applies the Chézy–Manning formula.

Polish workers have also obtained the most exact dates for the latest time that the former large meanders were active. Kosarski [113] places his first generation of paleomeanders in the climatic phases Bølling–Younger Dryas, about 13 000 to 10 000 B.P. Starkel and others [55] fix on a somewhat shorter interval, Allerød–Younger Dryas, or 12 000–10 000 B.P., for paleomeanders on the Vistula River, calculating that bankfull discharge then was equivalent to the 200-year flood of today. Starkel [154] puts the matter in a different way, by saying that the reduction of meander size of some Polish rivers corresponded to a reduction of channel-forming discharge by a factor of 5 to 10.

He also records increased, but less than maximal, channel-forming discharge in the intervals 8 500–7 500, 6 800–5 800, and 5 300–5 000 B.P. Knox [112], in a parallel study for the Driftless Area of Southwest Wisconsin, that excludes the largest paleomeanders of all, uses point bars to define former banktops, and the attributes of existing channels to define former 1.58-year floods. In the range studied, former channel forming discharges run from 10–15% larger to 20–30% smaller than those of today. High flows characterised the intervals 6 000–4 500, 3 000–2 000, and (briefly) shortly after 1 200 B.P. Low flows were common between about 8 000–6 500, 4 500–3 000, and 2 000–1 200 B.P. Whereas no evidence yet to hand opposes the idea that the latest operation of the largest former meanders, in areas then ice-free, belongs somewhere in the interval 13 000–10 000 B.P., the obvious discrepancies in the listings given for later times show, either that too little is yet known about the actual chronology, or that different regions have had different hydrologic histories.

The last episode of operation of the greatest paleomeanders was preceded by a history of downcutting, undoubtedly prolonged, that has produced the ingrown meandering valleys of existing landscapes. Since the dimensions of the ingrown bends are fully compatible with the wavelengths and channel sizes of the early Holocene, it follows that the discharges that shaped the valley bends were closely similar to, and may have been identical with, the discharges of the approximate interval 13 000–10 000 B.P. Nothing is known about fluctuations of discharge in earlier times. At the same time, the ingrowth of many meandering valleys was certainly spasmodic; and, when ingrowth began, many—and possibly all—of the channels concerned were more or less straight. Sinuosity developed during ingrowth. Nothing is known of the dating of the onset of ingrowth in areas beyond the reach of Pleistocene ice sheets.

**Lack of Meanders**

Some early studies of manifest underfits were handicapped by the failure of some
maps to portray existing stream meanders. Even when this matter is rectified by
appeal to air photographs or to accurate maps, however, it can be seen that not all
streams contained in meandering valleys possess meanders on their present channels. Artificial works are responsible in some cases, but by no means in all. The
most obvious apparent result is a very high wavelength/width ratio, say on the
order of 40/1: but the wavelength is that of the valley bends, the width that of the
existing channel. In all of the few cases that have so far been investigated, the
existing channels possess pool and riffle spaced appropriately to their widths, but
fail to effect side-to-side swing. To channels of this kind, the name *Osage-type
underfit* is applied, after the river for which the condition was first identified [55].
Since manifest underfits are recognisable on any accurate map, whereas Osage-type
underfits can only be identified by means of profiling of the bed—or, in favourable
cases, on air photographs at very low flow—it is not surprising that work on the
latter class has lagged behind that on the former. One study, on part of the English
river Severn [67], concludes that the former large stream was capable of deforming
both bed and banks, and was thus able to meander, while the existing Osage-type
underfit can deform its bed into the existing sequence of pool and riffle, but is not
capable of deforming its banks: hence the absence of stream meanders. Brown [24]
in a later study, wherein bed shear and bank shear are not separated, refers low
sinuosity on the lower Severn to low stream power.
    Far more work of this general kind is clearly needed, before any generalisations
about Osage-type underfits can be attempted. Modes of underfitness additional to
the manifest and Osage-type can be hypothesised, but have not yet been investigated.


**Climatic Change**

Alluvial meanders one-tenth the wavelength of paleomeanders, and existing channels one-tenth the width of their predecessor paleochannels, testify to conditions of
former greater surface wetness that can be matched from arid areas. Lake Dieri,
the precursor of the occasionally-flooded Lake Eyre in Australia, had a surface
area of more than 100 000 km$^2$. A formerly integrated drainage system in Western
Australia has been completely dismembered by a reduction of rainfall. The
ancestors of Lake Chad in Africa north of the equator, and of Lake Okavango in
the Kalahari, were vast bodies of standing water nourished by precipitation greater
than that of today. Thousands of other examples could be adduced, ranging down
to tiny pans where rims of gypsum testify to the former presence of standing ponds.
Calculations for numbers of sites combine to indicate that the former precipitation
required to produce the observed effects was about twice the annual average of the
present time—a value suggestively similar, perhaps, to that obtained from the
study of paleomeanders.

Deserts and semi-deserts, however, may preserve records fuller than those of midlatitude valleys. The Salt Lake and neighbouring pans in northwestern New South Wales reveal the following minimum sequence [60]: conditions dryer than those of today, with seif dunes active; conditions wetter than those of today, with 15 m of standing water in the Salt Lake, and erosion of seifs along the shores; conditions again drier than those of today, with renewed construction of seifs, some of them on the floors of the dried pans; and present conditions, with seifs of both generations stabilised and partly vegetated. If such fluctuations can occur within the 20 000 or so years locally available, one can only imagine the possible complexity of the climatic records of the main tropical deserts, which certainly date back at least to the Miocene.

For arid Australia as a whole, Bowler [20] records generally high lake levels before 25 000 B. P., with dunes at that time mainly stable; but then ensued the last major arid phase, which in the south of the continent peaked in 18 000–16 000 B.P., the stresses relaxing by 13 000 B.P. Whereas Bowler's results can be read to equate low-latitude aridity with high-latitude glaciation, Brook [23] for a considerable area of southern Africa finds cold and wet to very wet episodes coinciding with the Weichsel I, II, III glacial maxima and with one of the included interglacials, while his warm and dry interval from 14 000 to 9 700 B.P., which he equates with the Allerød and Bølling phases of the North European sequence, is in direct contrast to the moist conditions known to have obtained in Europe during part at least of the same interval. The seemingly attractive prospect of constructing parallel climatic chronologies for low (humid and arid), middle, and high latitudes still appears remote. The background assumption in taking such parallelism to be possible is that the proximate control was glaciation.

This matter is addressed by Street [157] in a helpful review of some 10 to 12 years of pertinent literature. Ocean cores from the eastern equatorial Atlantic show sand layers at perhaps 38–34 Ma, and certainly at 23–20, 13–12, and 3–2 Ma; piston cores show additional and younger layers. Here is the evidence for the antiquity of low-latitude deserts. Glacial maximum was associated with strengthened trade winds, which promoted increased upwelling, while temperatures in the main cold current systems were reduced. Air temperatures at the last glacial maximum* were lowered by 3 to 8 °C, enough to reduce the rain forests of Africa and South America to refugia. Maximum dust transport from northwest Africa coincided with low lake stands—in part a sign of reduced precipitation—in low latitudes. The interval 9 000–8 000 B.P. brought the greatest expansion of lacustrine conditions to Africa, the Middle East, and the American Southwest. Once again, the low-latitude and middle-latitude records of surface moisture fail to match. Although aridity in low latitudes has spread widely in the last 5 000 years, many equatorial lakes now stand higher than they did at the time of the last glacial maximum, about 18 000 B.P. Cores from the Niger delta show peak discharges in the intervals 22 000 ± 2 000 B.P., 13 000–11 800 B.P., and 11 500–4 000 B.P., in good agreement with the known high stands of the ancestral Lake Chad, but discordant with findings for the rivers of the humid midlatitudes.

A distinct optimist about the prospect of constructing general chronologies is Fairbridge, who, beginning with a model of eustatic changes of ocean level during the last 10 000 years [73], has subsequently discussed the effects of climatic change, during the same general period, upon some tropical geomorphic processes [75]. He concludes that the time of the last full-glacial stage of middle and high latitudes was typified in low latitudes by hyperaridity; that a postglacial pluvial occurred, as in middle latitudes, but somewhat later, with a peak for example at about 9 000 B.P.; and that subsequently there have occurred desiccation of lakes, readvances of deserts, fall of sea level, and truncation of coral reefs. He justly stresses the multidisciplinary character of the researches needed to provide solutions to the problems set by the evidence.

A warning that events for which a climatic trigger might be sought may have some other cause comes from the work of A.R.M. Young [185] on the upland swamps of a New South Wales plateau. Sandy organic sediment in shallow valleys has been subject to episodic erosion during the interval 17 000–300 B.P.; but there is no discernible pattern or clustering in the basal dates. Episodic clearance of sediment therefore fails to correlate from one swamp to another. Young, rejecting climatic shift as the initiating agent, concludes instead on natural wildfire.

That general atmospheric cooling which eventually resulted in Pleistocene glaciation appears to have been responsible for the retraction of what would now be called humid-tropical climates. One type of evidence is supplied by paleo-botany: analysis of the leaf margins of fossil flora makes it possible to establish limits of paleotemperature. The general picture to emerge for midlatitude areas is one of very considerable cooling, accompanied by major displacements of flora, from about the end of Eocene times onward. Geomorphologists enter the discussion in relation to relict deep weathering. Although a great deal of attention has been directed at the chemical makeup of the surface crusts associated with deep weathering—silcretic, ferritic, and ferralitic, these last two being often referred to as laterite—and also to physical character—massive, blocky, nodular—the essential fact for geomorphology is that deep chemical weathering, now confined to the humid tropics, formerly affected vast areas of middle latitudes [70].

Particularly with reference to silcretic crusts, some authors have sought to separate the deposition of surface material from the chemical rotting of the rock beneath. In Southwest Wisconsin, by contrast, deep rotting is directly associated with the formation of iron-rich crusts at two horizons, one on sandstone at the surface and the other at the underground junction between limestone above and sandstone below.

Deep chemical weathering destroys the radiometrically datable material of igneous rocks. The age of the youngest rocks to be deeply weathered thus gives a limiting lower date by stratigraphical means. Unfortunately, some of the formations are unfossiliferous or otherwise of uncertain age: such is the case in the Rhine and Mississippi Embayments. In southeast England, the London Clay of the Ypresian Stage of the Paleocene contains the fossil fruits of a flora of Indo-Malayan type. In northeast Ireland, pauses in the eruption of basalts allowed deep weathering to attack the new lavas; and a radiometric date of about 55 Ma is available from this area, supporting the fossil climatic evidence from the London Clay.

Reviewing paleobotanical evidence for the northern hemisphere, Wolfe [180] finds that a terminal Eocene event, during which mean annual temperatures declined by as much as 13 °C, took place in no more than one or two million years. Divergent latitudinal patterns since that time, which include a warming in the late to early Middle Miocene, fit very well with a significant decrease in the inclination of the earth's rotational axis. Meridional circulation of the atmosphere during the Eocene may have been ensured by an axial tilt in the range 5 to 10° of arc, but subsequently an initially slow and then rapid increase brought the tilt angle to 25–30° at the end of Eocene times. Superimposed on all this, according to Wolfe, is a temperature cycle of about 9.5 million years.

Reviewing Mesozoic climates, Hallam [91] describes these as far more equable than those of today. Shackleton and Boersma [149] use oxygen-isotope paleo-temperatures for the Eocene ocean, obtaining a latitudinal temperature gradient less than half the present, with surface temperatures at high latitudes about 10 °C. One consequence of the subsequent climatic deterioration is given by Donnelly [47] as a vastly increased rate of continental denudation, especially during the last 15 million years. The evidence consists in a sixfold increase in the rate of accumulation of $Al_2O_3$, mainly in hemipelagic rather than pelagic sites. Distributional facts militate against the choice of glaciation or tectonic movement as responsible agent: instead, Donnelly concludes on worldwide climatic deterioration, acting through the reduced power of vegetation to retain soil.

Because humid-tropical climates still exist, the retraction of deep weathering cannot fail to have been diachronous; but the terminal-Eocene event of Wolfe surely promoted the sudden cessation of deep weathering in huge areas of the earth's lands. On the scale of geologic time, that event constitutes a climatic step function.

Orbital and axial variations of the earth appear to lead naturally to the concept of cyclic change. Against this, Loubere and Moss [120] recognise step functions in late Pliocene climatic change and the onset of northern-hemisphere glaciation. Plate-tectonic movement seems incapable of explaining either the late Pliocene changes or the terminal-Eocene event: it works too slowly, and in some areas in the wrong direction. For instance, Australia, which contains evidence of deep weathering throughout its extent, was in Paleocene times united with Antarctica: the net equatorward movement subsequently has not prevented the retraction of deep weathering from leaving most Australian profiles in the relict state.

In the short term for which climatic statistics are available, a great deal of analytical damage seems to have been done by the practice of using running means (moving averages) in order to suppress statistical noise. As Slutzky and Yule pointed out as long ago as 1927 [153, 193], that practice can produce a misleading appearance of cyclic variation: the summation of random causes may be the source of undulatory processes. Curry, indeed, has approached climatic change as a random series [44]. As opposed to cyclic change, step-functional change is completely in accord with the crossing of geomorphic thresholds, and thus with sudden changes in systems behaviour.

## Neocatastrophism

At the far extreme from the concept of cyclic change stands the concept of neocatastrophism, so called because it constitutes a revival of the catastrophic thought popular among earth scientists in the early 19th century. Catastrophism was originally formulated by paleontologists, notably Cuvier, in the early 1800s, as a means of explaining what then seemed to be sudden extinctions. It spread into structural geology and geomorphology, leading to hypotheses of sudden crustal upheavals and fissuring. It fell into disrepute before the geological concept of uniformitarianism, and the mounting evidence that the surface history of the earth is measured in many millions of years.

Neocatastrophism also developed in paleontology, during the late 1950s and the early 1960s, when far more information was available about fossil numbers than had been a hundred and fifty years earlier. Neocatastrophism also spread into geology—sedimentology in particular—and into geomorphology, with attention turning to major and sudden events [61, 64]. Research into natural hazards has probably facilitated neocatastrophic thinking.

Extinctions sudden and extensive enough to be called catastrophic include those of shallow marine life at the end of the Permian Period, of dinosaurs and other life forms at about the Cretaceous/Tertiary (K/T) boundary, and of Pleistocene terrestrial megafauna. While these are the subject matter of paleontology, some of the hypotheses proposed to account for them have direct geomorphic implications.

The most intensely examined extinction is without doubt that of the dinosaurs, creatures that have caught the popular imagination. The first carefully reasoned hypothesis that their extinction resulted from the collision with the Earth of some other body appears to be that of Urey, who in 1973 proposed cometary collisions as agents of the termination of geologic periods [170]. In the same year, Christensen and others reported on an unusual concentration of platinum-group metals in Cretaceous/Tertiary clay in Denmark [35]; later, Alvarez and others reasoned from the presence of iridium spikes that terminal Cretaceous extinctions were due to collision of an asteroid [4]. This form of the collision hypothesis has become known well beyond the bounds of earth science; but Hoffman and Nitecki [97] conclude from a poll of opinion that, despite a perceptible cultural bias, the majority of workers are adverse to the asteroid hypothesis. On the other hand, there does appear to have occurred at the K/T boundary not only noteworthy extinction of fauna, but also of flora: Tschudy and Tschudy [169] describe the event as a profound ecological shock. The continuing debate is represented for instance by Bohor and others [19], who for a site in Wyoming identify at the K/T boundary an assemblage of shock-metamorphosed minerals, an iridium anomaly of 21 parts per billion, and the termination of numerous Late Cretaceous palynomorph species.

Taylor [164] supplies a variant on the asteroid-collision hypothesis, suggesting that the Moon originated in the collision with the Earth of a Mars-sized body, about 4.4 million years ago.

An anti-catastrophic view of the K/T boundary is expressed by Archibald and Clemens [7], who find that the transition to flora more typical of Paleocene than of

Cretaceous times began while the dinosaurs were still extant, that the severe floristic extinction used to mark the boundary is well above the level of the last dinosaur remains, and that the record as so far known indicates a gradual but in total very marked reduction of dinosaur taxa during, say, the last 10 million years of Cretaceous time. This whole matter is liable to be confused by stratigraphic difficulties, some of which have been discussed by Fastovsky and Dott [77].

The record of extinctions bears on geomorphology insofar as it leads to conclusions about climatic change. More directly relevant are events of sudden, massive, and rapid sedimentation, and events of enormously high stream discharge. Mullins and others [127] locate a submarine slide scar, at least 120 km long and up to 30 km wide, on the platform of west Florida, inferring sedimentation to have provided the trigger. For turbidites in the Lower Silurian of the Welsh Border, Benton and Gray [14] deduce the operation of storm surges and ebb traction currents, with a mean frequency or one of more events every several thousand years. Kastena and Cita [107] refer the deposition of an unusual grey marl in the basin topography of the abyssal Mediterranean to a single event, the collapse in 3 500 B.P. of the Santorini caldera: the resulting tsunami caused sediment to be shed from basin walls. On land, Porter and Orombelli [139] calculate that the catastrophic rockfall of September, 1717, on the Italian flank of the Mont Blanc massif, involved a volume of $15\text{--}20 \times 10^6$ m$^3$, and travel speeds of at least 125 km/hr: this event may be compared to falls described in literature cited earlier [61].

In a thoughtful paper that includes many useful references, Dott [49] prefers the term *episodic* to *catastrophic*, on the ground that the latter encourages the creationist-neocatastrophic cause: but the objection is unlikely to carry any weight outside the U.S.A., which alone seems to have developed creationism an an alternative to evolution. Dott does suggest *cataclysmic*, if a substitute term should be required; but such are the processes of language that the use of *catastrophic* might now seem too widespread to be eliminated. Among the single major events described by Dott is the Grand Banks turbidity current, which, initiated by a gigantic earthquake-triggered submarine slump, attained a flow volume as great as $10^{10}$ m$^3$. He also cites the fossil storm beaches in the Cambrian succession at Baraboo, Wisconsin: tropical hurricanes raised waves 6-8 m high which, beating against low islands of Precambrian quarzite, proved capable of rounding blocks as much as 1.5 m in diameter.

Like some other sedimentologists—in particular, those concerned with storm-generated deposits—Dott points out that Lyellian uniformitarianism may impose a subconscious abhorrence of unique events, discontinuities, and large deviations from 'average' conditions. The invocation of periodic cycles of sedimentation in an attempt to preserve uniformity is however counter to the actual geologic record. One is reminded at this juncture of Kuhn's comments on normal science and conceptual.boxes.

Dott is among the opponents of the asteroid-collision hyposthesis of extinctions at the K/T boundary, preferring instead to appeal to the new allergens of flowering plants as a means of killing off the dinosaurs. He further comments that, although the calculated recurrence interval of asteroid impact on the oceans is 30 million years, no suboceanic craters are yet known, nor is there known any evidence of the effects of waves with initial amplitudes of several kilometres. The average recurrence interval of meteorite impact capable of producing a crater 1 km or more in diameter is, for the earth's surface as a whole, is about 75 000 years, which can be translated into some 250 000 years for the land areas; but the proven number of such craters is only about one hundred. These could conceivably represent the result of 25 million years of meteoric bombardment, but only if all earlier craters had been somehow eliminated or concealed.

Investigators of cataclysmic floods of past times normally rely on equations that relate the calibre of material transported to flow velocity: if channel cross-section area can be measured, then flood discharge can be calculated. Costa [39] shows that checks on known floods indicate that the calibre method can lead to errors in calculated discharge of as much as 75 per cent either way. Even so, some impressive values appear. In addition to those mentioned elsewhere [64], Costa himself calculates, for the Biblical Deluge at Ur, a discharge of $1.5 \times 10^6$ m³/s. Kehew and Lord [109] arrive at discharges of up to $10^5$ m³/s at bankfull, for the rapidly draining spillway outlets of proglacial lakes on the Great Plains. Kehew [108] finds that, at such a rate of discharge through the Souris spillway, Glacier Lake Regina could have been drained in about one month. Jarrett and Malde [104] recalculate values for the late Pleistocene Bonneville Flood on the Snake River, Idaho, arriving at peak values more than twice the previous value, namely about 935 000 m³/s, with shear stress at 2 500 N/m² and unit stream power at 75 000 N/m/s. The total volume of water released becomes 4 700 km³, more than three times the previous estimate. These authors claim this to be the second largest known flood in the world, after the Lake Missoula outbreaks; but, as noticed, calculations for the Deluge at Ur and some of the Plains spillways run higher.

The outspills recognised so far as the greatest of all are those from the ice-dammed Lake Missoula in the Pacific Northwest of the U.S.A. Erosional forms and processes for this region, referable to discharges as great as $21.3 \times 10^6$ m³/s, and to velocities as great as 30 m/s, have been dealt with by Baker [9] and by Baker and Nummedal [10]. The chief difference of form between the depositional and erosional effects of the Missoula floods, and those to be observed in ordinary stream channels, is simply a matter of size, except that in some locations the cavitation process, rarely at work in ordinary channels, was effective.

When Bretz first described the terrain and proposed a flood hypothesis in 1923 [27], it was possible to think of a single unique event—something quite alien to the prevailing geological uniformitarianism of the day. Now that repeated outbreaks are inferred, uniqueness is no longer in question: Waitt [173] indeed has used more than 40 successive rhythmites, of 20 to 55 varves each, as evidence for colossal quasi-periodic outbursts, taking as the release mechanism the buoying up of the ice dam, and calculating that the lake could refill after each burst in three to seven decades. On this basis, the conditions prerequisite for an outburst could have been established 40 to 50 times, and possibly more, during the 2 000 to 2 500 years of damming during the last glacial maximum. The regime of outburst was clearly step-functional.

Neocatastrophism could be approached in another way, that is, through catastrophe theory, the main background reference on which is Thom [166]. An intuitive connection exists between one of more of the types of catastrophe with which Thom deals, and the crossing of geomorphic thresholds. Graf [83] considers that the cusp catastrophe model is likely to be the most useful to geomorphologists, since it is characterised by abrupt and smooth changes, divergent and bimodal behaviour, hysteresis, and stability of structure. Its advantages include a marriage of concepts of equilibrium and change, stability of change structure, and perspective. Its disadvantages include the difficulty of indentifying control factors, definition of energy functions, and generality; but, as Graf points out, the calculus-based mathematics so widely employed in the solid earth sciences is limited in its effectiveness in the context of abrupt transitions or discontinuities—in which geomorphologists are becoming increasingly interested.

## Rock Types and Earth Movements

Physical geology, as developed during the first half of the present century, dealt among other things with the influence of rock types and structure on the physical landscape. Because the influence of structure became entangled with discussions of the so-called erosion cycle, it lost a great deal of interest when the cycle concept became to some extent discredited: in any event, the network-geometric analysis of stream morphological systems was very largely independent of structure, while the emphasis placed on ultimate planation by advocates of the cycle concept caused structural influence to appear, in the long term, no more than transitory. Thus it became possible to adopt the view that landscape features not readily explicable in geologic terms were the proper material of geomorphology. Only for granitic and carbonate terrain, and to some extent in the case of arenite terrain, has any elaborate link been established between landform type on the one had and rock type on the other. Although moves toward a unification of approach are represented for instance by the work of Yatsu [184] and Mainguet [121], it is probably fair to say that attempts to relate rock types to climatic geomorphology, as this is understood on the European mainland, have failed to affect Anglophone geomorphologists. Granitic terrains pose problems of their own, since so many have been affected by deep weathering in the past, as for instance in Southwest England and the Channel Islands. A useful collection on the landscapes of arenites is that edited by Young and Nansen [192], in which Young [187] draws attention to the process of block gliding, whereby massive sandstone blocks move outward from a scarp at very low angles.

Whereas early geomorphology treated the effects of structures already in being—chiefly structures in sedimentary rocks such as faults and rifts, uniclinal structure, open folds, and close folds—some later work deals with structures that are still developing. The subdisciplinary name for the study concerned is *neotectonics*: according to the authors involved, treating of crustal movements from the Miocene onwards, from the Pliocene onwards, during Quaternary time, or during contemporary time. Except for glacioisostatic rebound, and for some work on the flexure of continental margins, most neotectonic studies belong to the second half of the present century. They are all process–response in character, working back for the most part from response to process. An obvious cause of their generally limited scope is that they require, in the ideal case, repeated geophysical survey. Another problem is that the time interval measured is liable to affect the measurement of rates of movement and change, both for tectonic processes and for geomorphic processes [81]. Among the neotectonic studies subsequent to those noticed by Mescherikov in the *Encyclopedia of Geomorphology* [123], are those of Adams [2] who discusses active tilting of the United States midcontinent; Burnett and Schumm [28] and Watson and others [146], who record, as responses to the uplift of the Monroe and Wiggins anticlines, deformed Quaternary terraces, floodplain and channel convexities, and variations of channel sinuosity, gradient, and depth; and Cronin [41], who for the emerged coastal plain of the southeastern U.S.A. uses paleontologic and radiometric evidence to arrive at net vertical uplift rates averaging 1 to 3 cm/1 000 yr, with a maximum of perhaps 5 to 10 cm/yr in places. These rates compare with subsidence rates of some 2 to 4 cm/1 000 yr since the Cretaceous, where the continental margin borders subsiding sedimentary troughs. Cronin's appeal to long-term flexuring in response to sediment loading offshore, although at odds with the history of Appalachian geosynclinal sedimentation, takes the discussion of continental flexuring back to the mid-century views of L.C. King.

It took more than 20 years for plate-tectonic theory to make a full impact on geology. The impact on geomorphology appears as yet slight. Nevertheless, a morphotectonics working group of the International Union, first proposed in 1978, was finally established shortly before the 1985 Binghamton geomorphology meeting which devoted itself to tectonic geomorphology [126]. In the conference collection, Summerfield [160] gives what he modestly calls a preliminary, qualitative account of plate tectonics and landscape development on the African continent. As is well known, Africa has been long claimed to have experienced successive episodes of landscape planation. Summerfield however finds significant discrepancies between the reputed cycle sequence and the sedimentary record at the continental margins: these discrepancies may be due either to errors in the landscape chronology, and/or to the complex responses of passive plate margins to changes in relative ocean level. The same author also considers continent-wide erosional cycles to be unlikely events. In the same collection, Bull [27] discusses correlations of global margine terraces, taking into consideration uplifts in the range 0.2 to 10.0 mm/yr, while Adams [3], studying the large-scale geomorphology of New Zealand's Southern Alps, obtains a mean erosion rate of $1.7 \pm 0.4$ mm/yr, against an upift rate of $1.5 \pm 0.1$ mm/yr; these values are sensibly enough identical to point the conclusion that the region's spiky mountains are time-independent, that is, in steady state.

While information on neotectonic movement is geomorphologically valuable, information on the absence of movement also has its worth. Nir and Eldar [132] use excavation data on wells, for the interval 3 100–700 B.P., to show that all ancient water marks almost merge with the present water table of the Mediterranean coastline of Israel. The indicated conclusion is that the region has been tectonically stable for the last three millenia. Until a very great deal more work has been done, however, it will not be known to what extent existing landscapes, and especially existing channels, have been and are being affected by neotectonic movement. In the broader view, a great deal needs to be learned about the relationship of geomorphology to plate-tectonic movement.


## Substitutes for Field Observation


If geomorphic time runs from about 25 to about 250 years, then it extends through about 1 to 10 working lifetimes. Even if a worker will monitor a site for 25 years, there can be no guarantee that monitoring will continue thereafter; but in some cases it may prove possible to retrieve information from the recent historical past. Some pertinent observations that depend on successive map surveys have been cited in connection with the development of channel sinuosity. Britain, with its original 1:10,560 Ordnance Survey Map of the 19th century, is well placed in this regard, although artificial works in some areas and indubitable errors of mapping in others limit the value of the evidence available.

The history of landscape photography is also now long enough to provide a supplement, or alternative, to map records. Graf [84, 85] has made effective use of photographs dating back to 1864–1871 in defining change, or absence of change, in stream channels receiving mine tailings or containing large boulders. Bryant [26] uses computer analysis of oblique photographs to reconstruct the former dimensions of a New South Wales beach in the interval 1895–1980: an unforeseen conclusion of his work is that above-average rainfall can promote beach erosion by raising the water table on the foreshore.

A dated landslide may supply a *terminus a quo* for future observations. Pearce and Watson [134] treat the effect of sliding triggered by a magnitude 7.7 earthquake in South Island, New Zealand, in 1929. The delivery of $400\,000\,m^3/km^2$ of sediment meant the burial of first- to third-order channels to depths of up to 10m over distances of 100 to 2 000 m; 50 years later, much of this sediment was still retained, for instance in fourth-order basins. The variation in recovery time promises to be so great that generalisation will be very difficult. At the other extreme, Costa [38] observes that the widening of stream channels in a Piedmont catchment in consequence of the Tropical Storm Agnes flooding of 1972, had been almost corrected in the space of a single year. We seem to know as little about recovery time as about thresholds.

In very many instances, the situations presented are those of a before-and-after character. The effects of the withdrawal of oil, water, or gas are usually not measured until subsidence becomes obvious, in which event a net measurement since starting date can be determined, but not variations in rate. Dolan and Goodell [46], in a comprehensive review, state that subsidence from natural causes (tectonic plus rise in sea level) can amount to 11.5 mm/yr, but that subsidence caused by extraction can reach 50 mm/yr. Many individual papers describe the effects produced on stream channels downstream of dams, but many also suffer from paying attention to too short a term. Andrews [6], measuring the effects of damming on the Green River in Colorado–Utah, where peak discharges were reduced and sediment discharge was cut by 50 per cent, finds that the 10 per cent reduction in bankfull width achieved in about a quarter of a century may by no means represent complete adjustment to changed conditions, and that a hundred years or more may be required to produce the required response. This and similar findings throw grave doubt on the rates of process measurements that can be made within the scope of an individual postgraduate project.

Some apparent surrogates for prolonged field observation are misleading. At one time there prevailed a perceptible fashion of measuring weathering rates on tombstones. It is true that the burial dates give starting points; but if the stones were emplaced fresh from the quarry, then they can be relied on to have weathered with undue rapidity, just as have the faces of certain Oxford colleges that were externally restored in Victorian times. Rates of erosion as measured by sedimentation in reservoirs are probably reliable, whatever allowance needs to be made for the effects of land use; but rates as measured by sedimentation in fishponds and small ornamental lakes are indubitably misleading, in view of the normal practice of periodic draining and cleaning.

Except where the alluvial fill of underfit streams has been drilled out, or where observations have been made downstream of dams, information on former channel dimensions is scanty. Two exceptions may however be mentioned. A naval survey party made a survey by sounding line, in the earliest 1870s, of the channel bed of the Hawkesbury River in New South Wales: the findings are still on record in the Admiralty archives [56]. Re-survey by sonar in 1969 [57] revealed, except for the disappearance of one meander pool that may have been obliterated by a landslide, no net change in pool depth, although individual pools had either shallowed or deepened. An approximate steady-state condition had been maintained for a century. By contrast, marked channel shrinkage has occurred on the eastern English Rivers Nene and Great Ouse, where discharge has been reduced by offtaking to water mills [59]; the original natural channels have adopted width/depth ratios appropriate to their adjusted sizes. Since, however, nothing is known of the date of the completion of adjustment, or of the dates of initial diversions, all that can be said is that adjustment was effected in some 1 000 years at maximum. Nothing firm can be learned here above recovery time. The same comment applies to the floodplain of the middle Nene, which today descends in steps, each riser being located at a mill dam and each tread resulting from alluviation on the upstream side.

## Some Broader Considerations

The above-mentioned possibility of variations in the rates of operation of geomorphic processes serves to introduce one of the reasons of geologic uncertainty perceived by Schumm [148], namely, singularity: this can be illustrated for example in short-term variation in the rate of meander shift. Divergence is shown when, for instance, increased precipitation produces contrasted effects in contrasted regions; it overlaps with the conflicting hypotheses of climatic change as the originator of arroyo development in the U.S. Southwest. Sensitivity is reflected in the triggering of major change by some minor input, as when slope failure occurs in an already established condition of incipient instability. Antecedent conditions and threshold conditions assume great importance in this connection. Schumm illustrates convergence by stream incision, which can result from relative lowering of the strandline or from climatic change; it may be added that the respective rates of response could be greatly different from one another, since, as has been seen for the Eastern Highlands of Australia, incision induced by uplift can be very slow indeed, whereas a significant increase in surface wetness and runoff throughout a catchment might be expected to promote a general and rapid reduction of channel slope. Not all authors however will follow Schumm in equating convergence with equifinality; to some, this latter term connotes the tendency of a natural system to revert, after disturbance, to an earlier condition. Convergence is perhaps most readily illustrated for minor features; arches may be cut by waves, or occasionally by meanders, or in some cases are due merely to the weathering of a spur on land, where the lower part is weaker than the upper. Small weather pits, the origin of which is not yet wholly clear, can closely simulate in shape the pebble–scoured potholes of a stream bed. Although the almost horizontal rock platforms of low-latitude coasts are in some respects unlike the abrasion-platforms of middle latitudes, they have deceived some observers. Pediments probably exemplify true convergence of response: it seems impossible to imagine that those of arid regions are shaped by the processes of soil creep that are at work on pediments of humid midlatitudes, while frost action is inferred to be the agent responsible for shaping cryopediments. The semilogarithmic profile form may be as fundamental to slope morphology as is the geometric relationship to network analysis or the normal curve to sedimentology.

Under the heading of complexity, Schumm points out that episodic behaviour may typify high-energy fluvial systems—an overlap, it would appear, with the concept of singularity, and also with the identification of step-functional flow regimes. Again, location can be significant, in that not all components in high-energy landscapes are functioning in phase: such is certainly the case where the magnitude–frequency relationships appropriate to events in channels are inappropriate to events on divides. Conclusions drawn about high-energy landscapes depend, therefore, on the part of the system studied: predictions from data for one location may not be useful elsewhere. Comments to this same effect have been offered by many fluvial morphologists, especially those dealing with meanders; and the difficulties of arriving at generalisations have been repeatedly stressed earlier.

Finally, Schumm takes scale problems to involve time, rate, and size. He tabulates nonevents, microevents, mesoevents, and megavents against a scale of time running from 1 day to $10^8$ years. At this latter mark, terrace formation is a nonevent. In 1 day, a megavent is a sudden slope failure: at $10^8$ years, it is mountain building. The past record reveals much about major features and slowly acting processes, but little about mesofeatures and nothing about microfeatures that become nonevents. The present reveals information on microfeatures and mesofeatures. The import of this seems to be that, at all times, a geomorphologist must be conscious of the scales on which he is attempting to work.

A somewhat different approach to the time scale by Gardner and others has already been noticed [81]. Penning–Rowsell and Townshend [135] discuss the effects of variation in spatial scale. For factors affecting stream slope in southeast England, these authors find that local slope variations are related to the size of bed material, whereas broader variations in reach slope are related more closely to discharge. Bed material size helps to explain variations in stream slope between, rather than within, lithologically homogeneous areas. At a regional scale, channel shape does not help to explain variations in local stream slope, but in a single reach is dominant. As investigations come to be progressively concentrated at a local level, variables which can change rapidly over short distances assume increasing importance.

A comparable study is that of Ebisemiju [71], who undertakes order-by-order principal components analysis of eight morphometric parameters—basin area, basin circularity ratio, total stream length, total drainage density, total channel stream frequency, average ground slope, relief ratio, and mainstream channel slope. His work shows that geomorphic investigation on Hortonian lines is very far from exhausted. Irrespective of spatial scale, striking similarities appear in the intercorrelation structures of basin morphology. At the same time, the strength of interaction among the variables in each correlation set varies with basin order: as order increases, the intensity of interaction among the planimetric variables increases, while that of slope attributes decreases. In total, stream network size, drainage texture, slope, and basin shape explain well over 90% of the variance for all the four basin orders studied.

**Toward the Future: A Fifth Geomorphic Revolution?**

The study of network geometry, the introduction of hydrologic data, and the application of quantitative techniques have been signalised as effecting revolutions in geomorphology. It has also been suggested that the application of systems theory may be on the way to effecting a fourth revolution. The network revolution may not yet have run its full course, as is demonstrated by network-based research still in progress, but in the theoretical sense is probably over. The implications of the hydraulic revolution are still being explored, not least on account of the time required to conduct the process–response studies that it has promoted, and also because there remains much to be done in the exploration of magnitude–frequency analysis. The quantitative revolution, highly successful in respect of the application of fairly simple statistical techniques and of data-processing, may still have far to go in the application of the methods of classical mathematics: for instance, Trofimov and Moskovkin [168], treating among other things the influence on slopes of downcutting, undercutting, and pediment–scree formation, set up evolutionary diffusion models by means beyond the capacity of most practising morphologists. Insofar as it is proving successful, the systems revolution appears to be working by a kind of osmosis: in some sense the systems, or gestalt, concept appears to be so self-evident that it cannot possibly give rise to objections, but, like the concept of scale, it does force a worker to define terms of operation.

Beyond all this, the fifth geomorphic revolution that may be in prospect depends on the application of modern advanced mathematics. As Culling [42, 43] states, quantification in geomorphology lags sadly behind mathematical progress. To quote him: 'It is as if we had opened some magic casement to find, between chance and necessity, one dimension and the next, a whole newworld of chaotic motions, strange attractors and periodic windows . . . we gaze upon an ocean of discovery between two continents once thought contiguous.' Accompanying the general discussions of Culling is the specific paper of Huggett [101], already an exponent of the applications of systems theory, on dissipative systems in geomorphology. Such of these systems as contain bifurcations have both deterministic and prob-abilistic elements. Since they have more than one solution, they call into question the notion of process domains; they possess varying degrees of susceptibility to the changes induced by fluctuations; and they exhibit varying responses to fluctuations on the local, regional, and global scales.

It may seem likely, therefore, that future advances in geomorphic analysis will move progressively further away from the classical methods of mechanics, and that such deterministic–probabilistic/random dilemmas as those posed by meandering channels will eventually be resolved.

### Acknowledgements

Figures 1–3 are redrawn, by permission, from three of A.N. Strahler's diagrams in Physical Geography, 1(1), 1980, and Figures 4–6 from three of D. Brunsden's diagrams in Geologia Applicata e Idrogeologica, 8 (1973). The permissions are gratefully acknowledged. I am also grateful for considerable correspondence with fellow-workers, and in particular for the personal communications listed in the Bibliography.

### References

1. Abrahams A D (1984) Water Res. Research 20: 161
2. Adams J (1980) Geology 8: 442
3. Adams J (1985) Large-scale tectonic geomorphology of the Southern Alps, New Zealand. In: Morisawa M, Hack J T (eds), Tectonic Geomorphology Allen and Unwin, Boston, p 105
4. Alvarez L W, Alvarez W, Asaro F, Michel H V (1980) Science 208: 1095
5. Anderson R S, Hallet B (1986) Bull. Geol. Soc. Amer. 97: 523
6. Andrews E D (1986) Bull. Geol. Soc. Amer. 97: 1012
7. Archibald J D, Clemens W A (1982) Amer. Scientist 70: 377
8. Bagnold R A (1960) Some aspects of the shape of river meanders, U. S. Geol. Survey Profl. Paper 282–E
9. Baker V R (1973) Erosional forms and processes for the catastrophic Pleistocene Missoula floods. In Morisawa M (ed) Fluvial geomorphology, State U. of New York, Binghamton, p 123

10. Baker V R, Nummedal D (1978) The Channeled Scabland, NASA, Washington, D.C.
11. Baker V R, Pickup G (1987) Bull. Geol. Soc. Amer. 98: 635
12. Barrett P J, Elston D P, Harwood D M, McKelvey B C, Webb P-N (1987) Geology 15: 634
13. Bassett J L, Ruhe R V (1973) Fluvial geomorphology in karst terrain. In: Morisawa M (ed) Fluvial geomorphology, State U. of New York, Binghamton
14. Benton M J, Gray D I (1981) J. Geol. Soc. Lond. 138: 675
15. Biedenharn D S, Raphelt N K, Montague C A (1984) Long-term stability of the Ouachita River. In: Elliott C M (ed) River meanders, Amer. Soc. Civil Engrs., New York, p 126
16. Bishop P, Young, R W, McDougall, I: J. Geol. 93, 455 (1985)
17. Blong R J (1984) Volcanic hazards: a sourcebook on the effects of eruptions. Academic, Sydney
18. Boardman J (ed) (1987) Periglacial processes and landforms in Britain and Ireland. Cambridge University Press, Cambridge
19. Bohor B F, Tiplehorn D M, Nicholls D J, Millard H J Jr (1987) Geology 15: 896
20. Bowler J M (1976) Earth-Sci. Rev. 12: 279
21. Bretz J H (1923) J. Geol. 31: 617
22. Brice J (1973) Meandering pattern on the White River in Indiana—an analysis. In Morisawa M (ed) Fluvial geomorphology, State U. of New York, Binghamton, p 179
23. Brook G S (1982) Catena 9: 343
24. Brown A G (1987) Z. Geomorph. N. F. 31: 293
25. Brunsden D (1973) Geol. Appl. e Idrogeol. 8: 185
26. Bryant E A (1985) Z. Geomorph. N. F. Suppl.–Bd. 57: 51
27. Bull W A (1985) Correlation of flights of marine terraces. In: Morisawa M (ed) Tectonic geomorphology, Allen and Unwin, Boston p 129
28. Burnet A W, Schumm S A (1983) Science 222: 49
29. Burton I (1963) Canad. Geog. 7: 151
30. Carson M A, Kirkby M J (1972) Hillslope form and process, Cambridge University Press, Cambridge
31. Carson M A, Petley T J (1970) Trans. Inst. Brit. Geog. 48: 71
32. Chorley R J (1962) Geomorphology and general systems theory. U.S. Geol. Survey Profl. Paper 500-B
33. Chorley R J (1966) The application of statistical methods to geomorphology; In: Dury G H (ed) Essays in geomorphology Heinemann, London, p 275
34. Chorley R J, Kennedy B A (1971) Physical geography, a systems approach, Prentice-Hall, London
35. Christensen L, Fregerslev S, Simonsen L (1973) Bull. Geol. Soc. Denmark 22: 193
36. Coates D R, Vitek J D (eds) (1980) Thresholds in geomorphology, Allen and Unwin, London
37. Cook P G (1986) Austral. Geog. 17: 133
38. Costa J E (1974) Water Res. Research 10: 106
39. Costa J E (1983) Bull. Geol. Soc. Amer. 94: 986
40. Cox N J (1987) Trans. Inst. Brit. Geog. N. S. 12: 250
41. Cronin T M (1981) Bull. Geol. Soc. Amer. 92: 812
42. Culling W E H (1986) Trans. Japan. Geomorph. Union 7: 221
43. Culling W E H (1987) Trans. Inst. Brit. Geog. N. S. 12: 57
44. Curry L (1962) Ann. Assoc. Amer. Geog. 52: 21
45. Dalrymple J B, Blong R J, Conacher A J (1968) Z. Geomorph. N. F. 12: 60
46. Dolan R, Goodell H G (1986) Amer. Scientist 74: 38
47. Donnelly T W (1982) Geology 10: 451
48. Donovan D T, Jones E J N (1979) J. Geol. Soc. Lond. 136: 187
49. Dott R H Jr (1983) J. Sed. Petrol. 53: 0005
50. Dury G H (1954) Amer. J. Sci. 252: 193
51. Dury G H (1956) Rev. Géomorph. Dynamique no. 11-12: 161
52. Dury G H (1964) Principles of underfit streams. U.S. Geol. Survey Profl. Paper 452-A
53. Dury G H (1964) Subsurface exploration and chronology of underfit streams, U.S. Geol. Survey Profl. Paper 452-B

54. Dury G H (1965) Theoretical implications of underfit streams, U.S. Geol. Survey Profl. Paper 452-C
55. Dury G H (1966) Austral. Geog. 10: 17
56. Dury G H (1967) Austral. Geog. Studies 5: 135
57. Dury G H (1970) Austral. Geog. Studies 8: 121
58. Dury G H (1970) Rivers and river terraces, Macmillan, London
59. Dury G H (1973) Magnitude–frequency analysis and channel morphometry. In: Morisawa M (ed) Fluvial geomorphology, State U. of New York, Binghamton, p 91
60. Dury G H (1973) Bull. Geol. Soc. Amer. 84: 363
61. Dury G H (1975) An. Acad. Brasil. Cîenc. 47 (suplemento): 137
62. Dury G H (1976) J. Hydrol. 30: 219
63. Dury G H (1977) Peak flow, low flow, and geomorphic dominance. In: Gregory K J (ed) River channel changes, Wiley, Chichester, p 61
64. Dury G H (1980) Prog. Phys. Geog. 4: 391
65. Dury G H (1981) An Introduction to Environmental Systems, Heinemann, London
66. Dury G H (1982) Catena 9: 379
67. Dury G H (1983) Osage-type underfitness on the River Severn near Shrewsbury, Shropshire, England. In: Gregory K J (ed) Background to palaeohydrology, Wiley, Chichester, p 399
68. Dury G H (1984) Bankfull discharge through pool and riffle. In: Elliott C M (ed) River meandering Amer. Soc. Civil Engrs, New York, p 545
69. Dury G H (1985) Earth Surf. Procs. Landforms 10: 205
70. Dury G H, Habermann G M (1978) Australian silcretes and some northern-hemisphere correlatives. In: Langford-Smith T (ed) Silcrete in Australia, Dept. of Geog., U. of New England, Armidale, p 223
71. Ebisemiju F S (1985) Catena 12: 261
72. Embleton C, Brunsden D, Jones D K G (eds) (1978) Geomorphology, Present problems and future prospects, Oxford University Press, London
73. Fairbridge R W (1961) Eustatic changes in sea level. In: Ahrens L H, Press F, Rankama K, Runcorn S K (eds) Physics and chemistry of the Earth, 4, p 99
74. Fairbridge R W (ed) (1968) The encyclopedia of geomorphology, Reinhold, New York
75. Fairbridge R W (1976) Quatern. Res. 6: 529
76. Fairbridge R W, Hillaire–Marcel C (1977) Nature 268: 413
77. Fastovsky D E, Dott R H Jr (1986) Geology 14: 279
78. Ferguson R I (1973) Earth Surf. Procs. 9: 1079
79. Flood R D, Damuth J E (1987) Bull. Geol. Soc. Amer. 98: 728
80. Ford D C (1973) Canad. J. Earth Sci. 10: 366
81. Gardner T W, Jorgensen D W, Schumm C, Lemieux C R (1987) Geology 15: 254
82. Goudie A (ed) (1981) Geomorphological techniques, Allen and Unwin, London
83. Graf W L (1979) Catastrophe theory as a model for change in fluvial systems. In: Rhodes D P, Williams G P (eds) Adjustment of the fluvial system, Kendall–Hunt, Dubuqe, Iowa, p 13
84. Graf W L (1979) J. Geol. 87: 533
85. Graf W L (1979) Ann. Assoc. Amer. Geog. 69: 262
86. Graf W L (1983) The arroyo problem—palaeohydrology and palaohydraulics. In: Gregory K J (ed) Background to palaeohydrology, Wiley, Chichester, p 279
87. Graf W L (1986) Progr. Phys. Geog. 7: 97
88. Graf W L (1987) Bull. Geol. Soc. Amer. 99: 261
89. Gregory K J, Walling D E (1973) Drainage basin form and process, Wiley, New York
90. Grove J M (1987) Geog. J. 153: 351
91. Hallam A (1985) J. Geol. Soc. Lond. 142: 433
92. Hancock J M, Kauffmann E C (1979) J. Geol. soc. Lond. 136: 175
93. Haq B U, Hardenbol J, Vail P J (1987) Science 235: 1156
94. Hickin E J (1978) Canad. J. Earth Sci. 15: 1833
95. Hickin E J (1983) Spec. Pub. Inst. Assoc. Sed. 6: 61
96. Hickin E J, Nanson G C (1975) Bull. Geol. Soc. Amer. 86: 487

97. Hoffman A, Nitecki M H (1985) Geology 13: 884
98. Horton R E (1932) Trans. Amer. Geophys. Union 13: 350
99. Horton R E (1945) Bull. Geol. Soc. Amer. 56: 275
100. Huggett R J (1985) Earth surface systems, Springer, Berlin Heidelberg New York
101. Huggett R J (1987) *pers. comm* and (in press) Earth Surf. Procs. Landforms
102. Imbrie J (1985) J. Geol. Soc. Lond. 142: 417
103. James W R, Krumbein W C (1969) J. Geol. 77: 544
104. Jarrett R D, Malde H E (1987) Bull. Geol. Soc. Amer. 99: 127
105. Johnson J G, Klapper G, Saunders C A (1985) Bull. Geol. Soc. Amer. 96: 567
106. Karcz I (1980) Thermodynamic approach to geomorphic thresholds. In: Coates D R, Vitek J D (eds) Thresholds in geomorphology Allen and Unwin, London, p 209
107. Kasteno K A, Cita M B (1981) Bull. Geol. Soc. Amer. 92: 845
108. Kehew A E (1982) Bull. Geol. Soc. Amer. 93: 1051
109. Kehew A E, Lord M L (1986) Bull. Geol. Soc. Amer. 97: 192
110. Keller E A, Brookes A (1984) Consideration of meandering in channelization projects...In: Elliott C M (ed) River meandering Amer. Soc. Civil Engrs., New York, p 384
111. Kirkby F (1987) *pers. comm.*
112. Knox J C (1985) Quatern. Res. 23: 287
113. Kozarski S (1983) River channel adjustment to climatic change. In: Gregory K J (ed) Background to palaeohydrology, Wiley, Chichester, p 355
114. Kuhn T S (1970) The structure of scientific revolutions (2nd. edn.), University of Chicago Press, Chicago 1970
115. Kukla G J (1977) Earth-Sci. Rev. 13: 307
116. Laczay L A (1977) Channel pattern changes of Hungarian rivers. In: Gregory K J (ed) River Channel Changes, Wiley, Chichester, p 154
117. Langbein W B, Leopold L B (1964) Amer. J. Sci. 262: 782
118. Langbein W B, Leopold L B (1966) River meanders—Theory of minimum variance. U. S. Geol. Survey Profl. Paper 422-H
119. Leopold L B, Wolman M G, Miller J P (1964) Fluvial processes in geomorphology, Freeman, San Francisco
120. Loubere P, Mors K (1986) Bull. Geol. Soc. Amer. 97: 818
121. Mainguet M (1982) Le Modélé des Grès, Inst. Geog. Nat., Paris
122. Melton M A (1958) J. Geol. 66: 442
123. Mescherikov Y A (1968) Neotectonics. In: Fairbridge R W (ed) Encyclopedia of geomorphology Reinhold, New York, p 768
124. Morisawa M (1985) Development of quantitative geomorphology. Geol. Soc. Amer. Centennial Spec. vol 1
125. Morisawa M (1985) Topologic properties of delta distributary networks. In: Woldenburg M (ed) Models in geomorphology, Allen and Unwin, London p 239
126. Morisawa M, Hack J T (eds) (1985) Tectonic geomorphology, Allen and Unwin, Boston
127. Mullins H T, Gradulski A F, Hine A C (1986) Geology 14: 167
128. Nanson G C (1986) Austral. Geog. 17: 87
129. Nanson G C, Young R W (1981) Z. Geomorph. N. F. 25: 332
130. Neolson J, Kocurek G (1987) Bull. Geol. Soc. Amer. 99: 177
131. Nickling W G (ed) (1986) Aeolian geomorphology, Allen and Unwin, Boston
132. Nir Y, Eldar I (1987) Geology 15: 3
133. Oldfield F (1983) The role of magnetic studies in palaeohydrology. In: Gregory K J (ed) Background to Palaeohydrology, Wiley, Chichester, p 141
134. Pearce A J, Watson A J (1986) Geology 14: 52
135. Penning–Rowsell E C, Townshend J R G (1978) Trans. Inst. Brit. Geog. new ser. 3: 395
136. Philbrick S (1970) Bull. Geol. Soc. Amer. 81: 3723
137. Pilgrim A T, Puvaneswarn P, Conacher A J (1986) Catena 13: 169
138. Pitty A F (ed) (1985) Themes in geomorphology, Croom Helm, London
139. Porter S C, Orombelli G (1980) Z. Geomorph. N. F. 24: 200
140. Rambsbottom W H C (1979) J. Geol. Soc. Lond. 136: 147

141. Rapp A (1960) Geog. Annaler 42: 71
142. Reid I, Frostick L E (1986) Earth Surf. Procs. Landforms 11: 143
143. Rohdenburg H (1982) Geomorphologisch–bodenstratigraphischer Vergleich Zwischen dem Nor-
     dostbrasilianischen Trockengebiet und Immerfeucht–Tropischen Gebieten Südbrasiliens. In:
     Ahnert F Rohdenburg H, Semmel A (eds) Catena Supp. 2, p 73
144. Ross C A, Ross R J P (1985) Geology 13: 194
145. Rotnicki K (1983) Modelling past discharge of meandering rivers. In: Gregory K J (ed)
     Background to palaeohydrology, Wiley, Chichester, p 321
146. Scheidegger A E (1958) Principles of geodynamics, Springer Berlin Heidelberg New York
147. Scheidegger A E (1961) Theoretical geomorphology, Springer, Berlin, Heidelberg New York
148. Schumm S A (1985) Trans. Japan. Geomorph. Union 6: 1
149. Shackleton N, Boersma A (1981) J. Geol. Soc. Lond. 138: 153
150. Short A D, Wright L D (1984) Morphodynamics of high-energy beaches in Australia. In: Thom B
     G (ed) Coastal geomorphology in Australia, Academic, Sydney, p 43
151. Shreve B L (1966) J. Geol. 74: 17
152. Simons M (1962) Trans. Inst. Brit. Geog. 32: 1
153. Slutzky E (1927) The summation of random causes as the source of cyclic processes. In: The
     Conjecture Institute, (ed) Problems of economic conditions, 3 (1), Moscow (in Russian)
154. Starkel L (1983) The reflection of hydrologic change in the fluvial environment. In: Gregory K J
     (ed) Background to Palaeohydrology, Wiley, Chichester, p 213
155. Starkel L, Klimek K, Mamakowa K, Niedziakowska E (1982) The Wisoka River valley in the
     Carpathian foreland. In: Starkel L (ed) Evolution of the Vistula river Valley during the last 15 000
     years, Polish Acad, Sci. Geog. Studies Spec. Issue 1, Pt. I, Warsaw, p 41
156. Strahler A N (1952) Bull. Geol. Soc. Amer. 63: 923
157. Strahler A N (1952) Bull. Geol. Soc. Amer. 63: 1117
158. Strahler A N (1980) Phys. Geog. 1: 1
159. Street F A (1981) Prog. Phys. Geog. 5: 157
160. Summerfield H A (1985) Plate tectonics and landscape development on the African continent. In:
     Morisawa M, Hack J T (eds) Tectonic geomorphology, Allen and Unwin, Boston, p 27
161. Tanner W F (1974) Shale Shaker 24: 128
162. Tanner W F (1973) The littoral power gradient and shoreline changes. In: Coates D R (ed) Coastal
     Geomorphology, State U. of New York, Binghamton, p 43
163. Tanner W F (1986) So-called bedload, the hydrograph, and sediment transport. In: Wang S Y,
     Shen H N, Ding L Z (eds) River sedimentation School of Engineering, U. of Mississippi, p 375
164. Taylor S H (1987) Amer. Scientist 75: 468
165. Terzaghi K (1943) Theoretical soil mechanics, Wiley, New York
166. Thom R (1975) Structural stability and morphogenesis: an outline of a general theory of models,
     Benjamin, Reading
167. Thompson B (c. 1895) MS Notebooks, Northampton County Museum
168. Trofimov A M, Moskovin V M (1984) Earth Surf. Procs. Landforms 9: 435
169. Tschudy R H, Tschudy B D (1986) Geology 14: 667
170. Urey H C (1973) Nature 242: 32
171. Vail P R, Mitchum R M, Thompson S III (1977) Seismic stratigraphy and global changes of sea
     level, Pt. 4. In: Payton C E (ed) Seismic stratigraphy—applications to hydrocabron deposition,
     Amer. Assoc. Petrol. Geol. Mem. 26, p 83
172. Von Bertalanffy L (1950) Science 111: 23
173. Waitt R B Jr (1985) Bull. Geol. Soc. Amer. 96: 1271
174. Walling D E, Webb B W (1983) Patterns of sediment yield. In: Gregory K J (ed) Background to
     palaeohydrology, Wiley, Chichester, p 69
175. Wang J, Derbyshire E (1987) Geog. J. 153: 59
176. Watson C C, Schumm S A, Harvey M D (1984) Neotectonic effects on river pattern. In: Elliott C
     M (ed) River meandering, Amer. Soc. Civil Engrs., New York, p 55
177. Williams G P (1978) Water Res. Research 14: 1141
178. Williams G P (1984) Paleohydrologic equations for rivers. In: Costa J E, Fleischer P J (eds)
     Development and applications of geomorphology, Springer, Berlin Heidelberg New york, p 343

179. Winkley B R (1984) Factors influencing crossing (riffle) depth. In: Elliott C M (ed) River meandering, Amer, Soc. Civil Engrs., New York, p 343
180. Wolfe J A (1978) Amer. Scientist 66: 694
181. Wolman M G (1955) The natural channel of Brandywine Creek, Pennsylvania. U.S. Geol. Survey Profl. Paper 271
182. Wolman M G, Miller J P (1960) J. Geol. 68: 54
183. Wood A (1942) Proc. Geol. Assoc. 53: 337
184. Yatsu E (1966) Rock control in geomorphology, Sozosha, Tokyo
185. Young A R M (1986) Z. Geomorph. N. F. 30: 317
186. Young R W (1983) J. Geol. 91: 221
187. Young R W (1983) Block gliding in the sandstones of the southern Sydney Basin. In: Young R W, Nanson G C (eds), Aspects of Australian sandstone landscapes Austral. N. Z. Geomorph Gp. Spec. Pub. 1, p 37
188. Young R W (1985) Z. Geomorph. Suppl.–Bd. 55: 81
189. Young R W (1986) Austral. Geog. 71: 91
190. Young R W (1987) J. Geol. 95: 205
191. Young R W, McDougall I (1985) Austral, J. Earth Sci. 32: 323
192. Young R W, Nanson G C (eds) (1983) Aspects of Australian sandstone landscapes. Austral. N.Z. Geomorph. Gp. Spec. Pub. 1
193. Yule G (1927) Phil. Trans. Roy. Soc. A, 126: 267

# Global Transport Processes in the Atmosphere

*James R. Holton*

Department of Atmospheric Sciences, University of Washington Seattle, WA 98195, USA

## Summary

Mechanisms for global transport of atmospheric trace constituents are discussed in the context of the meteorological processes that are responsible for irreversible transport. The characteristics of the circulations associated with global transport are first described. Next the nature of transport processes in the troposphere and the stratosphere are discussed, and processes accounting for exchange of trace species between the troposphere and the stratosphere are reviewed. Finally, various types of transport models are discussed including simple box models, one dimensional models, two dimensional models and three dimensional models.


## Introduction

It has been realized for many years that local and regional air quality are often substantially affected by human activities. Recently, there has been a growing appreciation that industrial and agricultural activities are contributing to significant modifications of the atmosphere not only in the vicinity of pollution sources, but on a global scale. The best known example of such a global effect is the increasing concentration of carbon dioxide caused primarily by fossil fuel burning. Carbon dioxide and other trace gases that are efficient absorbers in the infrared are important contributors to the so-called greenhouse effect. The increases projected in the global concentrations of such substances will lead to an increase in the infrared opacity of the atmosphere. The resulting increase in the flux of radiation from the atmosphere to the surface is expected to have a significant effect on surface temperatures in the next century.

Although carbon dioxide is chemically inert in the troposphere and stratosphere and is fairly uniformly mixed, a number of other radiatively important trace gases such as methane, ozone, and the chlorofluorocarbons undergo chemical transformations in the atmosphere that can be strongly dependent on location and time. The concentrations of such tracers thus tend to become highly variable. Unless the time scale for chemical change is very short compared to time scales characteristic of meteorological systems, their distributions are strongly affected by atmospheric motions. Thus, to understand the implications of trends in globally significant trace species it is necessary to consider not only their sources, sinks, and chemical transformations within the atmosphere, but also their transport by the winds. It is this last aspect of atmospheric tracers that is the focus of this chapter.


### Transport Processes

Atmospheric transport processes are conveniently divided between those processes that involve bulk motions of the atmosphere, referred to as "advection" by meteorologists, and those processes that may be characterized as turbulent, or diffusive in nature. In the case of point sources, such as power plant plumes, the distinction is quite clear; advection moves the center of mass of the plume along the direction of the average wind, while turbulent diffusion disperses the plume in the plane orthogonal to the average wind. On a global scale distinction between advective and diffusive processes is not always clear. Since the atmosphere is characterized by spatially and temporally varying motions with a wide range of scales, there is no obvious physical separation between "mean" and "turbulent" motions. In practice, those transport processes that are explicitly resolved by the particular observational network or transport model being utilized are often regarded as the advective motions, while the remaining unresolved motions are assumed to be diffusive. The effects of the unresolved motions can then be parameterized in terms of the mean motions by assuming that tracer fluxes by the unresolved motions are proportional to the gradient of the resolved tracer distribution. However, this approach is not always physically justified. A major problem in the modeling of global transport is an accurate representation of the contribution of the unresolved "diffusive" motions to the total transport.

## Global Circulation of the Atmosphere

In order to understand what is meant by global atmospheric transport it is first necessary to discuss the nature of the general circulation of the atmosphere. That is, we must understand the global distribution and variability of the winds that accomplish the transport. In characterizing the circulation of the atmosphere it is useful to begin by separating the atmosphere into various layers based on the vertical profile of the average temperature (Fig. 1). For our purposes it is only necessary to consider the two lowest layers. These are the troposphere, which varies in depth from less than 8 km near the poles to about 15 km in the equatorial zone, and the stratosphere, which extends from the top of the troposphere to about the 50 km level. These two layers are separated by the tropopause. In the troposphere the temperature tends to decrease with height at a mean rate of about 6 K km$^{-1}$,

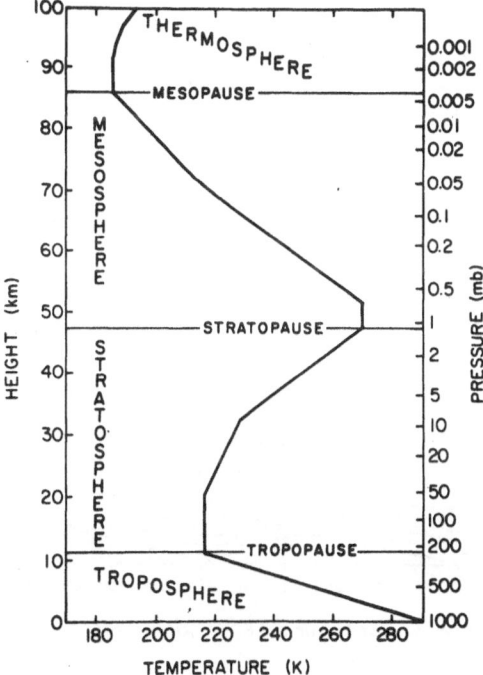

Fig. 1. Standard midlatitude temperature profile showing the various layers of the atmosphere

while the lower stratosphere tends to be isothermal. In the middle and upper stratosphere there is a gradual increase of temperature with height. Thus, the tropopause marks a change in the vertical component of the temperature gradient. The differing temperature profiles in the troposphere and stratosphere are responsible for large differences in the effectiveness of vertical mixing. In the troposphere convection tends to promote rapid vertical mixing, while in the stratosphere vertical mixing is very weak.

The earth is a rapidly rotating planet; the circulation of the atmosphere is strongly affected by this rotation. Viewed globally the mean winds tend to blow parallel to latitude circles with characteristic speeds in the range of $10\text{--}40\,\mathrm{m\,s}^{-1}$ in the troposphere, so that trace constituents with isolated sources can become distributed around the globe within a few weeks. Thus, the lowest order role of global transport is to tend to homogenize tracer distributions in the longitudinal direction. The rate at which this occurs is, however, strongly dependent on the location of the source relative to the flow configuration.

The importance of transport processes for understanding climate can be illustrated by considering the solstice season zonally averaged temperature profile shown in Fig. 2. As the figure illustrates, the temperature decreases from the equator toward both poles in the troposphere, while in the upper stratosphere there is a monotonic increase from the winter pole to the summer pole. Careful calculations of the radiative equilibrium temperature, that is the temperature distribution that would occur in the absence of atmospheric motions, so that each vertical column of atmosphere would establish an equilibrium between solar heating and infrared

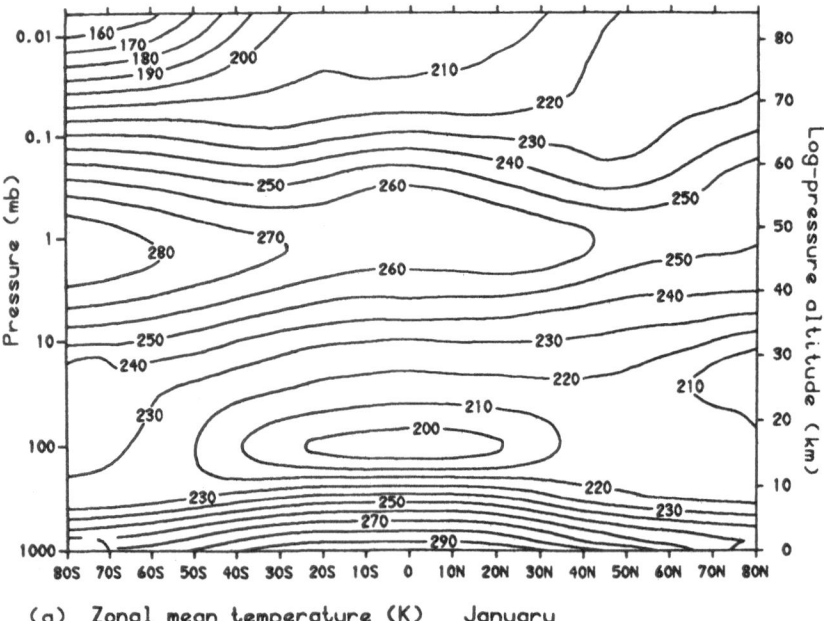

(a)   Zonal mean temperature (K)    January

**Fig. 2.** Meridional cross-section of zonally averaged temperature (K) for January, courtesy of J.J. Barnett and M. Corney, Department of Atmospheric Physics, Oxford University

cooling, indicate that the observed pole to equator temperature difference in the troposphere is only about 1/2 of that which would be expected at radiative equilibrium. Thus, there must be substantial heat transfer from the equatorial to the polar regions in order to explain the observed climate. Although perhaps as much as 50% of the required transport may be due to the ocean currents, much of it must represent heat transported by atmospheric motions. These atmospheric heat fluxes are associated with meridional motions. The distribution and variability of such motions is a primary aspect of climate variability. Since processes that transport heat poleward may also transport trace constituents, the nature of the general circulation determines the characteristics of global transport of trace constituents.

### The Circulation of the Troposphere

Because the pressure and mass distributions in the atmosphere are in hydrostatic balance (except locally in the presence of intense small scale circulations such as convective clouds), pressure is a monotonic function of altitude in the atmosphere. Furthermore, pressure, rather than height, may be used as the independent vertical coordinate (isobaric coordinates). Meteorological variables are conventionally mapped on surfaces of constant pressure, so that the isobaric coordinate system is a natural one for analysis and modeling. In the isobaric system the horizontal distribution of height at constant pressure plays the same role as the pressure distribution at constant height in height coordinates. Thus, the forcing of the winds by pressure gradient forces can be deduced from the gradient of height at constant pressure.

In Fig. 3 we show the contours of the average geopotential height of the 50 kPa pressure level (about 5 km altitude) for Northern Hemisphere winter. The Coriolis force due to the rotation of the earth, which acts to the right of the horizontal velocity in the Northern Hemisphere, and the pressure gradient force, which is directed down the height gradient, tend to be approximately in balance. (The velocity field for which this balance is exact is called the *geostrophic* wind; see, for example, [2].) Thus, the large scale flow is nearly parallel to the lines of constant geopotential height, with height increasing to the right of an observer facing downwind, and the highest flow velocities occuring where the contours are tightly packed together (i.e., where there is a strong gradient of geopotential height). Notice that the flow tends to be concentrated in a wavy "jetstream" extending around the globe, but with the wind speed varying substantially in longitude. This longitudinal variability is thought to be related to the influence of topography and continent-ocean heating differences. In the Southern Hemisphere there is a similar jetstream structure, but less longitudinal dependence because of the weaker influence of mountains and continents.

If the entire atmospheric flow field consisted only of the geostrophic winds flowing parallel to the mean height contours (such as those shown in Fig. 3), there would be no net transport of a conservative tracer in the meridional direction; the mean geostrophic wind would simply carry the tracer around the globe, oscillating it alternately poleward and equatorward parallel to the wavy height contours.

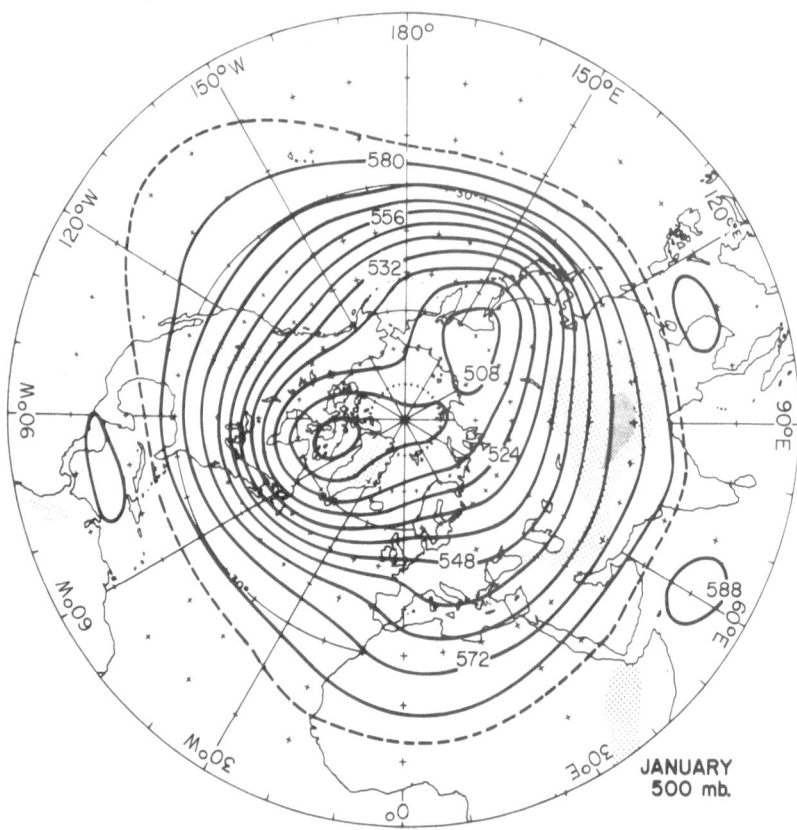

**Fig. 3.** Contours of the topography of the 500 mb surface for Northern Hemisphere January mean (units: decameters). From Palmen and Newton [1]

The actual flow field is more complex. The flow is not exactly in geostrophic balance; there are substantial ageostrophic components of the motion. These components, directed across the height contours, can produce net transport of substances in the meridional plane even in a steady state flow regime. The primary such circulation is the tropical Hadley circulation, which is characterized by upward flow in the equatorial region, poleward flow in the upper troposphere, and subsidence at higher latitudes. Although the strength of the Hadley circulation is longitudinally dependent, the dominant aspect is that of an overturning circulation in the meridional plane. When pictured in a time and longitudinally averaged sense, the upward branch of the Hadley circulation appears to consist of a gentle equatorial upwelling of a few centimeters per second. The actual upward transport, however, is known to be concentrated within the convectively active regions, where giant cumulonimbus clouds transport heat, moisture, and other tracers directly from the surface layer to the upper troposphere in less than an hour. Outside of the convectively active regions the vertical motion in the tropics is actually observed to be downward. Thus, the Hadley circulation should be viewed as a statistical rather than a physical entity.

The nature of this average circulation, particularly its meridional scale, is dependent critically on the type of longitudinal average used to derive the "average". When the average is taken at constant pressure or constant height along latitude circles (the convenional Eulerian mean), the circulation in midlatitudes is actually reversed from that of the tropical Hadley cell; the flow in the upper troposphere is equatorward and in the lower troposphere is poleward. If, on the other hand, the longitudinal average is computed along isentropic surfaces, uniform poleward directed flow is found in the upper troposphere (see Fig. 4). This sort of dramatic difference in the mean Hadley circulation reflects the fact that the actual flow contains components that have strong longitudinal asymmetries. The vertical and meridional structure of such eddies determines their contribution to the

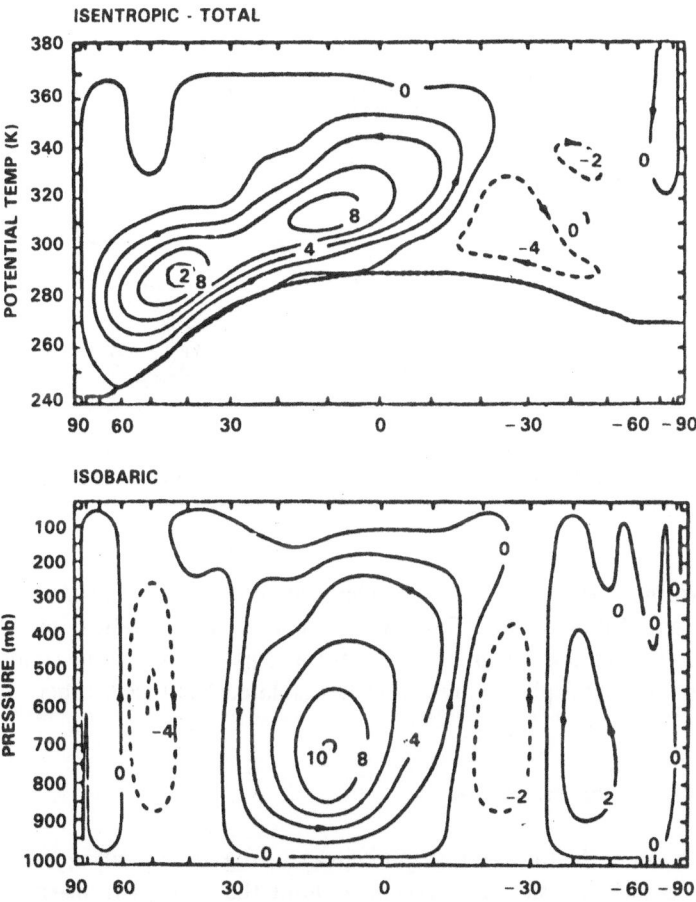

**Fig. 4.** The zonally averaged mean meridional mass streamfunction for Northern Hemisphere winter conditions in isentropic coordinates (top) and isobaric coordinates (bottom). Arrows show direction of circulation. From [3]

"mean" that results from a particular averaging process. Thus, it is necessary to be very cautious in drawing conclusions concerning global tracer transport from the structure of a longitudinally averaged flow field.

Whatever the form of averaging, the Hadley circulation alone cannot explain all the major features of meridional transport. Observations indicate that, outside the tropical zone, transport is not due solely to mean meridional motions, but is strongly influenced by rapid mixing due to transient cyclonic and anticyclonic circulations (referred to as synoptic systems by meteorologists) and planetary scale wave motions. The effects of such motions in dispersively transporting tracers is shown graphically by the model study of Fig. 5. Marked air parcels that originate from a small region in space are quickly stretched out into narrow wavy filaments by the eddy motions. Thus, transport in the troposphere within each hemisphere involves not only slow advective meridional drift, but rapid quasi-horizontal mixing as well.

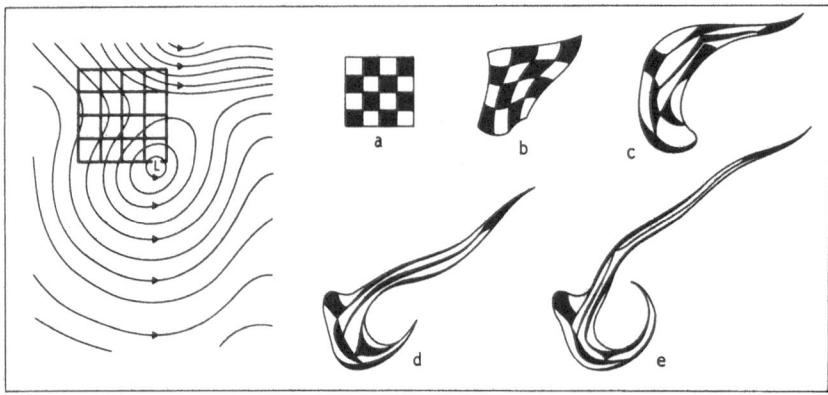

**Fig. 5.** The evolution of the shape of a set of marked fluid parcels initially forming a "checkerboard" pattern under the influence of a flow field (shown on the left) typical of that occurring in large scale atmospheric eddies. From Welander [10]

In the tropics the situation is somewhat different because the eddy motions are primarily equatorially trapped modes, which do not produce net meridional transport. Thus, to some degree the equator acts as a barrier to meridional mixing so that the two hemispheres are somewhat isolated from each other.

**The Circulation of the Stratosphere**

The circulation of the lower stratosphere may be regarded as an upward extension of the tropospheric circulation, except without the vertical transfer due to convective clouds. The synoptic scale transient eddies decay rapidly with height above the tropopause, so that throughout most of the stratosphere the motion systems are planetary in scale. Above its lowest layers the global scale circulation of the stratosphere is dominated by a mean flow that is westerly (from the west) in the winter hemisphere, and easterly in the summer hemisphere. During the equinoxes the flow tends to be westerly, but weaker than during the winter.

In the summer hemisphere the flow is nearly parallel to latitude circles. Large scale eddy motions have very small amplitudes, and meridional transport is mainly due to slow ageostrophic meridional drift. In the winter (especially in the Northern Hemisphere) the flow is disturbed by planetary scale waves of zonal wavenumbers 1 and 2 (1 and 2 wavelengths around a latitude circle). Thus, the geostrophic flow in the winter hemisphere departs significantly from axial symmetry. The mean flow distribution for the Northern Hemisphere features a cyclonic polar vortex that is distorted by the presence of a semipermanent anticyclonic disturbance, the Aleutian high. At times this feature may amplify dramatically over a short span of time, and produce rapid meridional transport. The heat transport associated with such events leads to the "sudden stratospheric" warmings of the North polar regions [4]. In a major warming the temperature at the 1 kPa level (about 31 km) may increase by as much as 40–60° C in less than a week. Such major warmings occur only in the Northern Hemisphere, and only once every other year or so. Warmings of a less intense nature occur throughout the winter in both hemispheres, and lead to episodic pulses in the meridional transport. In the Southern Hemisphere the polar vortex region appears to be isolated from such events throughout the winter so that there is very little transport of heat or tracers into the vortex from lower latitudes. The resulting chemical isolation of the region appears to be important for occurrence of the Antarctic springtime ozone hole [5, 6, 7].

### Transport Theory

The study of global transport involves the motion of atmospheric tracers, defined as chemical or dynamical quantities that label fluid parcels. Chemical tracers consist of minor atmospheric species that have significant spatial variability in the atmosphere. Dynamical tracers (potential temperature and potential vorticity) are properties of the flow field that are conserved following the motion under certain conditions. These also can be useful for interpretation of transport.

The crucial dynamical tracer for our purposes is the *potential temperature*, defined as the temperature that a parcel of air would acquire if it were adiabatically compressed from its ambient pressure to the standard pressure of 100 kPa. Because the atmosphere is stably stratified, potential temperature increases monotonically with height (slowly in the troposphere and rapidly in the stratosphere); like pressure, it can be used as an independent vertical coordinate. The motion referred to a system with latitude, longitude and potential temperature as the independent variables is generally called an *isentropic* coordinate system, since the logarithm of potential temperature is proportional to the entropy per unit mass. Except in regions of active precipitation, large scale motions in the troposphere and stratosphere are quasi-adiabatic. That is, the rate of accession of heat through diabatic processes (such as latent heating due to condensation of water vapor) is slow compared to the temperature change due to advection. Because a parcel moving adiabatically remains on a surface of constant potential temperature, its motion is two-dimensional when viewed in isentropic coordinates. Thus, isentropic co-ordinates are particularly useful in discussing the theory of transport.

The other commonly used dynamical tracer, potential vorticity, is basically a fluid dynamical analogue of angular momentum in a solid body. Potential vorticity is conserved for adiabatic, frictionless flow [8]. Its distribution can be determined from conventional meteorological observations. Thus, it can be used to assess transport in circumstances where there are no observations of chemical tracers.

### Theory of Transport in the Meridional Plane

In the previous section we noted that it is often useful to divide the flow between mean and disturbance portions. In transport modeling the mean of a quantity such as the mixing ratio q is usually a longitudinal average (denoted by an overbar) while the disturbance or eddy motion is the longitudinally asymmetric component (denoted by a prime):

$$q = \bar{q} + q'$$

Zonal averaging is extensively used in modeling of the stratospheric ozone layer. It is also used in some tropospheric problems. However, in discussing tropospheric transport a time mean is sometimes more appropriate.

As mentioned earlier, the most common longitudinal averaging is the *Eulerian mean* in which the averaging is taken over longitude at a fixed latitude, pressure (or altitude), and time. In transport problems this type of averaging, which is based on mathematical rather than physical principles, is often not an appropriate way of separating eddy and mean transport processes. For transport, the physical pro-cesses are most clearly revealed by use of a *Lagrangian mean* in which the averaging is taken over a fixed ring of fluid particles as they move about in space. However, the Lagrangian mean has proved very difficult to implement for realistic atmospheric flows [9]. A more useful framework, at least for the stratosphere, is based on using the isentropic coordinates described above, which mimic some aspects of the Lagrangian mean.

In transport studies the concentration of a tracer is generally expressed in terms of its mixing ratio rather than its density, since in the absence of sources or sinks, the mixing ratio is conserved following the motion. Mixing ratio can be expressed as a mass mixing ratio (tracer density divided by air density) or a volume mixing ratio (tracer number concentration divided by number concentration of air). In the formulation of transport theory given here it is convenient to use mass mixing ratios. The continuity equation for a tracer mixing ratio can be expressed as

$$\frac{d\chi}{dt} = S \tag{1}$$

which simply states that the rate of change of $\chi$ following the motion is equal to the net effect of sources and sinks (S). For a conserved tracer (S = 0) the transport problem in principle can be solved simply by following each air parcel as it moves about in space and time since the amount of tracer contained should remain constant. In practice this procedure is only practical over a short range of time since the trajectories of air parcels are extremely complex. Parcels that are initially cubical volumes may quickly become highly deformed as shown in Fig. 5. This would make it difficult to keep track of such parcels for a long period even if the motion field were perfectly resolved. But since there are always motions on scales smaller than resolved by the observations (or model) there will inevitably be some exchange of air between the resolved parcel and its environment. Hence, for practical purposes exact conservation of atmospheric tracers following the resolved motion does not apply.

In studies of global transport the Eulerian approach is usually used, but generally the problem is simplified by averaging over the longitudinal coordinate. This approach is justified since for most globally distributed tracers the meridional and vertical gradients are of far more significance than are the longitudinal gradients. In order to understand the significance of zonal averaging, we constrast the longitudinally averaged form of the tracer continuity equation for the isobaric and isentropic coordinate systems.

We will here use a form of isobaric coordinates in which the vertical coordinate is defined to be proportional to the log of pressure:

$$z = -H \ln(p/p_s)$$

where H is a mean scale height, p is pressure and $p_s$ is a standard pressure. If we let $p_s = 100$ kPa and $H = 7$ km then z is approximately equal to the geometric height. Mass conservation for dry air can be expressed in this system by the continuity equation

$$\nabla \cdot V + \frac{1}{\rho_0} \frac{\partial}{\partial z}(\rho_0 w) = 0 \tag{2}$$

where $V = (u, v)$ is the horizontal velocity, $w = dz/dt$ is the "vertical velocity" in the log-pressure coordinates, and $\rho_0$ is a reference density profile proportional to $\exp(-z/H)$.

Eq. (1) together with the continuity equation (2) can be combined to yield the flux form of the tracer continuity equation:

$$\frac{\partial \chi}{\partial t} = -\left[ \nabla \cdot (V\chi) + \frac{1}{\rho_0} \frac{\partial}{\partial z}(\rho_0 w \chi) \right] + S. \tag{3}$$

Here, the local rate of change of $\chi$ (for a volume element fixed in space) is given by the convergence of the flux of $\chi$ into the volume element plus the net difference between chemical sources and sinks.

The conventional Eulerian zonal mean of equation (3) is given by

$$\frac{\partial \bar{\chi}}{\partial t} = -\left[ \bar{v}\frac{\partial \bar{\chi}}{\partial y} + \bar{w}\frac{\partial \bar{\chi}}{\partial z} \right] - \left[ \frac{\partial}{\partial y}(\overline{\chi' v'}) + \frac{1}{\rho_0} \frac{\partial}{\partial z}(\rho_0 \overline{w' \chi'}) \right] + S. \tag{4}$$

$$[A] \qquad\qquad\qquad [B]$$

Here, the transport of $\chi$ in the meridional plane is clearly split into two portions, a "mean" advective transport by the Eulerian mean meridional motion (term A) and an "eddy transport" (term B). In early transport models this eddy term was generally parameterized as an eddy diffusion; thus, it represented a dispersion or mixing process. This approach was not, however, satisfactory as can easily be seen by examining the mean meridional motion field as shown in Fig. 4. Clearly the reversed mean meridional cell at midlatitudes is not consistent with observed poleward and downward transport in that region. Apparently the eddy flux term must be responsible for a net advective effect; it is not simply dispersive. That is, it is not simply down the gradient of the mean mixing ratio (see [11] and refs.). Because the eddy flux as defined by the conventional Eulerian mean itself contains a part of the advective transport, this averaging does not provide a clear separation of advective and diffusive processes.

An alternative averaging scheme, which provides a simpler conceptual view is the Lagrangian mean introduced by Andrews and McIntyre [12]. The reason for the simplicity of the Lagrangian mean budget equation is illustrated in Fig. 6. By definition the Lagrangian mean mixing ratio is just the average along a material tube at a reference height and latitude as shown in Fig. 6a. After a small time t the material tube is displaced to a new reference center of mass (Fig. 6b). But is distorted by the eddy motions. If, however, $S=0$ and exchange between the tube and its environment is neglected, the tube conserves its value of $\chi$ so that the Lagrangian mean evaluated at the new reference latitude and height is equal to the original value. The continuity equation for the Lagrangian mean tracer mixing ratio has the simple form

$$\frac{\partial \bar{\chi}^L}{\partial t} = -\bar{v}^L \frac{\partial \bar{\chi}^L}{\partial y} - \bar{w}^L \frac{\partial \bar{\chi}^L}{\partial z}. \tag{5}$$

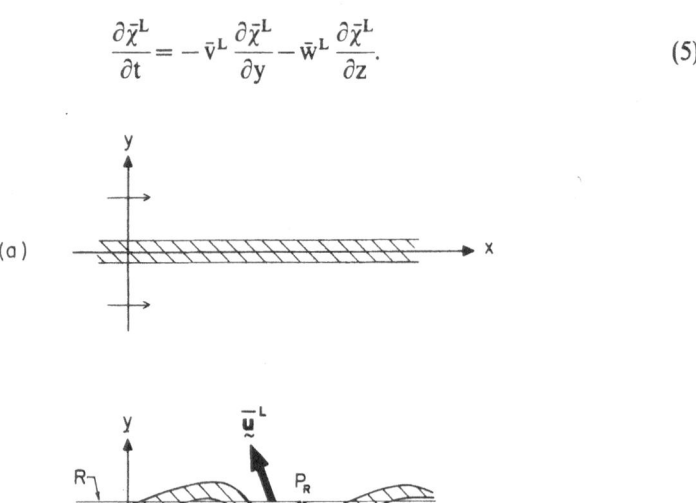

**Fig. 6.** Schematic illustration of the definitions of the Lagrangian-mean velocity $u^L$ and the parcel displacement x. The material tube is shown hatched. From [8]

Although the Lagrangian mean provides the simplest conceptual description of mean transport (eddy terms do not explicitly appear), its practical application is rendered difficult for a number of reasons. Firstly, it is very difficult to determine the Lagrangian mean flow ($\bar{v}^L$, $\bar{w}^L$) that must be used to advect the material tubes, although for some purposes the diabatic circulation discussed below may be a reasonable approximation. More importantly, unless the total flow field has a high degree of longitudinal symmetry, the Lagrangian tubes quickly become highly distorted. Just as in the case of parcels in the 3-dimensional trajectory analysis, the tubes become difficult to follow, and although formally they may define the center of mass, in practice parcels associated with a given tube may be located at widely differing latitudes and heights.

An alternative formulation that retains many of the desirable conservation features of the Lagrangian mean, but does not have its practical difficulties, is the isentropic coordinate mean. In this case, as in the conventional Eulerian mean, the average is taken along a latitude circle. But now the average is parallel to a surface of constant potential temperature rather than an isobaric surface. Since in the presence of eddies with significant temperature variations, the potential temperature surfaces may vary substantially in altitude around a latitude circle, it is clear that this sort of averaging may produce much different results than the isobaric average. The resulting equation can be written with some approximations as follows:

$$\frac{\partial \bar{\chi}}{\partial t} = -\left[ \bar{v}^* \frac{\partial \bar{\chi}}{\partial y} + \bar{Q}^* \frac{\partial \bar{\chi}}{\partial \theta} \right] - \frac{1}{\bar{\sigma}} \frac{\partial}{\partial y} (\bar{\sigma} \overline{v'\chi'}) + \bar{S}^*. \tag{6}$$

Here $\theta$ is the potential temperature, $Q = d\theta/dt$ is the diabatic heating rate, which plays the role of "vertical velocity" in the isentropic coordinate system, and $\sigma = -\partial p/\partial \theta$ is the "density" in the isentropic system. (Note that $\sigma \Delta \theta/g$, where $g$ is gravity, is the mass per unit horizontal area contained in a column of thickness $\Delta \theta$; (see [8]). In (6) the variables ($\bar{v}^*$, $\bar{Q}^*$) designate mass weighted longitudinal averages along isentropic surfaces. In this case the advective transport given by the first term on the right in (6) is due to the diabatic circulation so that as suggested by Fig. 4 the gross characteristics of observed transport in the meridional plane can be represented. Other formulations of the 2-dimensional transport problem are reviewed in Andrews et al. [8].

In (6) it is assumed that the eddies are nearly adiabatic so that the eddy flux is parallel to the isentropic surface. Only the mean diabatic heating is retained. The eddy flux in the second term on the right is then usually parameterized by letting

$$\bar{\sigma} \overline{v'\chi'} = -\bar{\sigma} K_{yy} \partial \bar{\chi}/\partial y \tag{7}$$

where $K_{yy}$ is a diffusion coefficient for quasi-isentropic eddy mixing that must be specified.

## Quasi-horizontal Dispersive Transport

The zonally averaged isentropic coordinates provide a useful framework for analysis of the bulk advective motions in the meridional plane that are associated with the diabatic circulation. However, in addition to the mean transport circulations, there is a very important contribution to transport in the form of sporadic rapid meridional transport due to large scale wave motions that may interact in a nonlinear fashion to produce dispersion of trace chemicals. It is this sort of process that we attempt to approximate in the diffusion parameterization of (7). Such dispersion is associated with irreversible wave breaking processes caused by the meridional displacements of material contours by the velocity perturbations associated with synoptic scale and planetary scale wave disturbances. The eddy dispersion process is especially effective in changing the distributions of tracers that have strong gradients (i.e., those with point or line sources). For example, radioactive debris from atmospheric nuclear tests were observed to spread rapidly along the isentropies in the lower stratosphere [13].

### Vertical Transport

In the planetary boundary layer, consisting of about the first km of atmosphere above the surface of the earth, the vertical transport depends strongly on the character of the potential temperature stratification. For convectively unstable layers in which the potential temperature decreases with height in the boundary layer, vertical transport is extremely rapid and the boundary layer tends to be well mixed. When there is stable stratification of the boundary layer it is reasonable to approximate boundary layer transport by an eddy diffusion process. Vertical transport in the free troposphere cannot, however, be represented purely in terms of a simple diffusive or advective process. Rather, it is necessary to consider the transport effects of convective clouds. As will be discussed below in connection with tropical stratosphere–troposphere exchange, cumulonimbus clouds provide powerful elements for rapid vertical transport of air from the boundary layer to the upper troposphere. Such transport clearly cannot be represented as a diffusion proportional to the mean gradient. Rather, it appears necessary to utilize a cumulus cloud model to parameterize the vertical transport of transient tracers by cumulus clouds. Models of this type have been developed by Russell and Lerner [14] and Gidel [15]. Some results from Gidel's work are shown in Fig. 7. It is clear from the figure that transport by convective clouds is a very effective mechanism for transport of trace chemicals from the surface to the detraining region in the upper troposphere, and that this transport can not be approximated by a simple diffusion model.

### Transport in the Troposphere

In recent years the chemistry of the global troposphere has been the subject of a number of observational and theoretical studies. For most problems of global tropospheric chemistry it is necessary to consider not only the sources, sinks, and chemical reactions of various species, but also the direct and indirect effects of transport. Even short lived species, such as the hydroxyl radical, which are in local

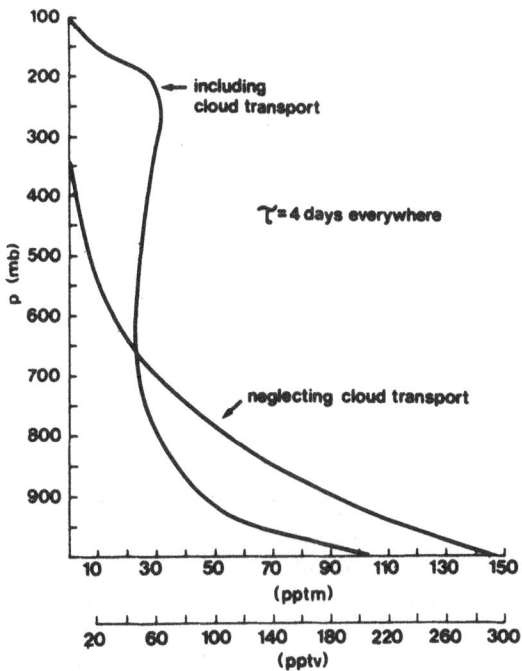

**Fig. 7.** Vertical profiles of tropospheric tracers with short lifetimes with and without the influence of vertical transport by convective clouds. From [14]

chemical balance, have budgets influenced by transport, since their sources (e.g., ozone, water vapor, ultraviolet radiation modulated by clouds) and sinks (e.g., carbon monoxide, methane, nonmethane hydrocarbons) are themselves transport controlled.

In the previous section we have seen that the details of transport can be quite complex. However, for many purposes it is only necessary to know the extent to which transport acts to homogenize particular tracers in the global troposphere. To examine this problem it is useful to consider two types of tracers: The first are tracers with very long tropospheric lifetimes (order of years); the second are tracers that undergo chemical transformations in the troposphere, but with lifetimes of order of a few weeks. For both these types transport from source to sink regions is crucial for establishing the global budgets. But for the first type transport is able to establish rather uniform tropospheric distributions, while for the second type transport is crucial in determining the distribution, but is not fast enough to eliminate substantial concentration differences between source and sink regions.

**Tropospheric Tracers**

There are a number of trace gases that are released at the earth's surface through natural processes or human activities. Those with sufficiently long lifetimes may be transported far from their sources. For such tracers local concentrations are dependent on global transport. Some tropospheric trace species, such as CO, are photochemically active in the troposphere, but have sufficiently long lifetimes (order of months) to be transported globally. Others, such as various chemicals involved in the sulfur cycle, tend to be removed by wet or dry deposition processes within several days so that they are subject to regional rather than global transport.

Among the various tropospheric tracers those whose observed distributions are most useful in providing information on global transport are several of the long-lived gases which are source gases for stratospheric ozone destruction. These include the synthetic chlorofluorocarbons (CFCs), $CFCl_3$ and $CF_2Cl_2$, and the naturally occurring gases nitrous oxide and methane.

### Interhemispheric Differences

Observations of tropospheric trace gases are limited mainly to sampling at a small number of stations, mostly in the Northern Hemisphere. Thus, global distributions are generally not well known. For tracers with lifetimes in excess of several months the winds tend to reduce the longitudinal variability so that zonal means provide good estimates except in the vicinity of concentrated sources (such as the industrial regions of Europe and North America). In a few cases systematic observations have been made that can be used to derive some information on differences between the hemispheres. Obviously, trace species that have industrial sources, or biological sources that are associated with agricultural activities, will tend to have their maximum concentrations in the Northern Hemisphere, since such sources are concentrated in the North.

The contrast between Northern and Southern Hemisphere concentrations is particularly striking for the CFCs, which are produced mainly in the North, and have concentrations increasing by several percent per year. Fig. 8 shows trends in the mixing ratios for $CFCl_3$ measured at several remote sights ranging from high Northern latitudes to Southern midlatitudes. Comparisons of the atmospheric burden with the known source strength suggest an atmospheric lifetime of about 75 years for this gas (Prinn et al. [16]). This is consistent with a stratospheric sink by photochemical destruction.

Fig. 8 also indicates that although the percentage difference in concentration between the two hemispheres is decreasing in time, the absolute concentration difference has remained nearly constant, reflecting the continued dominance of the Northern Hemispheric sources.

Nitrous oxide is produced primarily by natural processes that are distributed in both hemispheres. It has a lifetime of about 150 years in the atmosphere and a tropospheric concentration of about 300 ppb. Although there is evidence of a small secular increase (0.25% per year), the long lifetime causes $N_2O$ to be rather well mixed in the troposphere. However, the imbalance between sources and sinks implied by the secular increase indicates that transport is important in accounting for the global budget.

Methane, which has substantial sources and sinks in both hemispheres, has an atmospheric lifetime of about 10 years. The concentration is known to be increasing

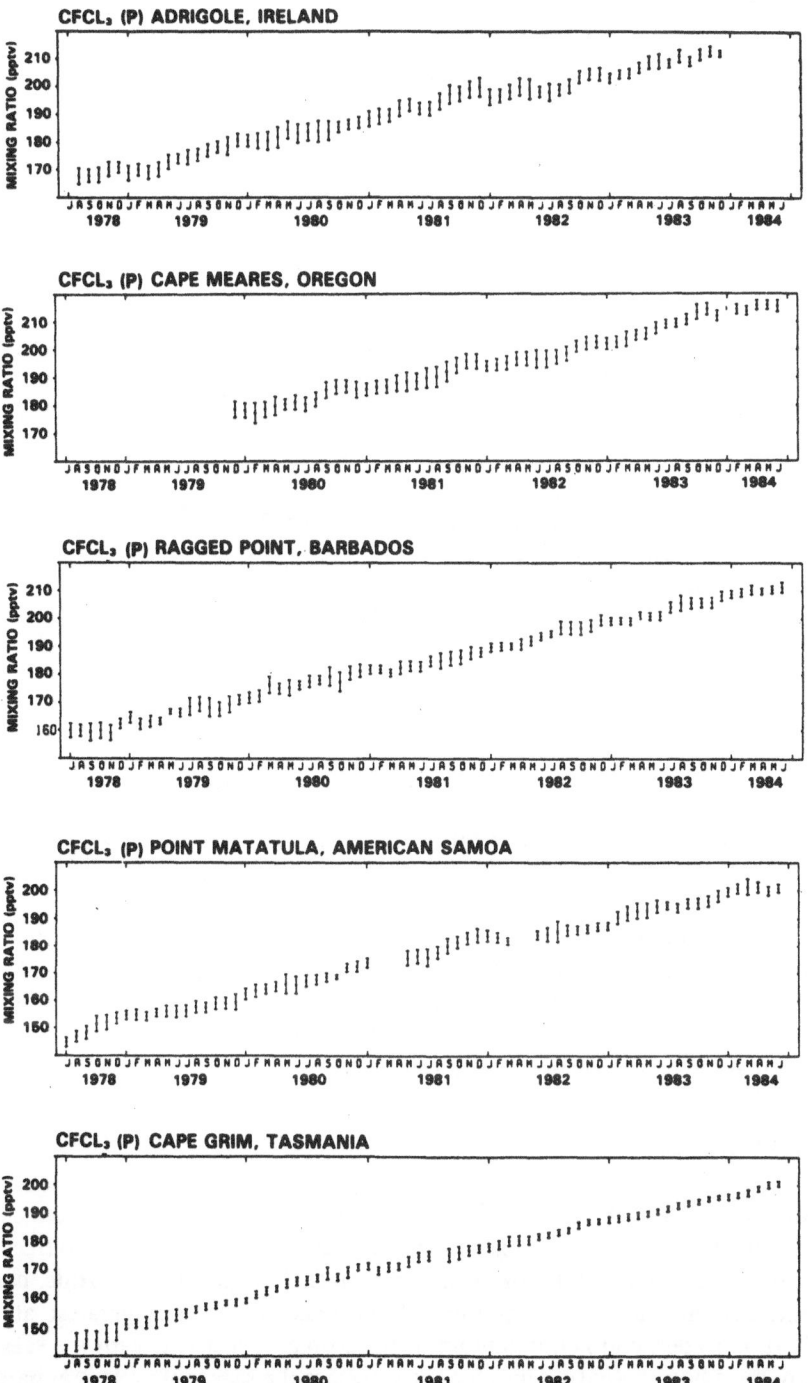

**Fig. 8.** Monthly mean surface concentrations of CFCl$_3$ at stations ranging from midlatitude Northern Hemisphere to midlatitude Southern Hemisphere. Units are volume mixing ratio in parts per trillion (ppt). After [3]

at a rate of about 1 percent per year, and there is evidence that the total concentration has doubled in historic times (Craig and Chou [17]). A variety of biospheric sources contribute to the methane burden [3]. Although the relative importance of the various sources is not well known, it is clear from the latitudinal distribution (Fig. 9) that the source strength must be greatest in the Northern Hemisphere.

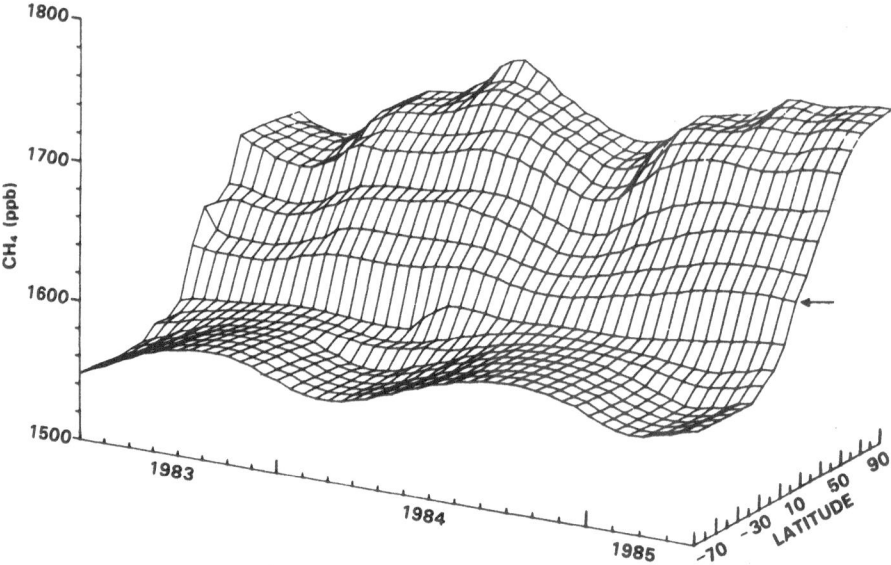

**Fig. 9.** Zonally averaged distribution of $CH_4$ mixing ratio in the lower troposphere for May 1983 through April 1985. Arrow indicates equator. From [3]

This sort of interhemispheric difference is also seen in the much shorter lived species, carbon monoxide (Fig. 10) and tropospheric ozone. For these, and other tracers it appears that contributions due to human activities—particularly fossil fuel burning—are now so widespread that enhanced concentrations are observable on a global, or at least hemispheric, scale.

**Seasonal Variations**

Seasonal changes in the tropospheric concentration of a long lived tracer can be caused either by seasonal changes in the strength or distribution of sources and sinks, or by seasonal oscillation in the transport. For carbon monoxide, carbon dioxide, and methane the amplitude of the seasonal cycle is greatest at high northern latitudes, and is a minimum in the tropics, consistent with the seasonal cycle in biospheric activity. Thus, for these species the seasonal cycle is apparently not strongly influenced by transport. The maintenance of a strong interhemispheric difference in the seasonal cycle also indicates that the time scale for exchange between the hemispheres is longer than the seasonal time scale.

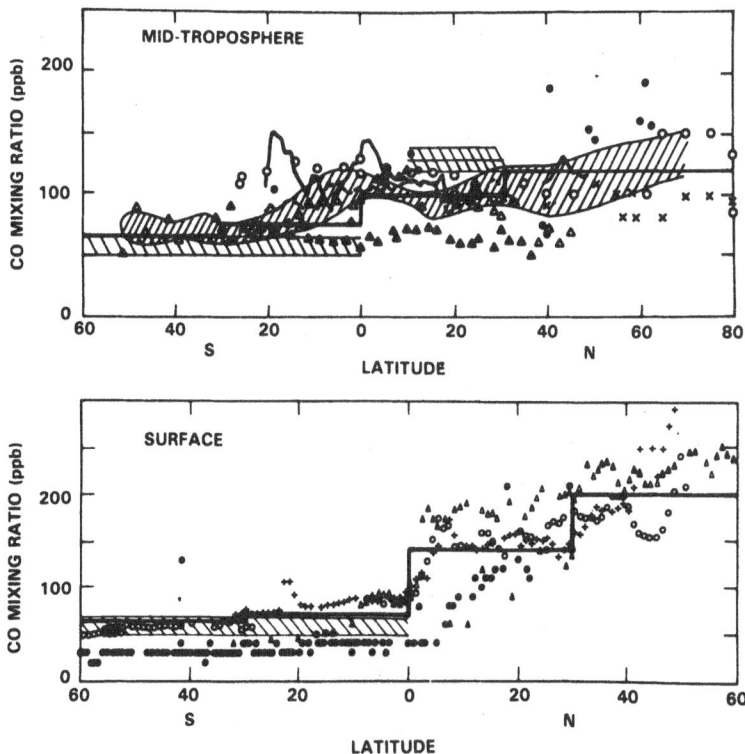

**Fig. 10.** Atmospheric CO measurements at the surface as a function of latitude for 1967 through 1978. From [18]

For ozone the situation is more complex. The traditional view is that tropospheric ozone is controlled by transport of ozone from the stratosphere, particularly in association with developing cyclonic storms. This process would produce a springtime maximum since the concentration of ozone in the lower stratosphere is greatest in late winter and spring, and troposphere–stratosphere exchange is also strongest at that time. This transport controlled view is supported by data for the upper troposphere (Fig. 11). However, in the middle and lower troposphere, as shown in Fig. 11, there is a maximum in the Northern Hemisphere summer. Logan [19] has argued that this maximum is produced by photochemical sources associated with continental scale pollution in the industrialized north. It appears that in the budget of tropospheric ozone, transport from the stratosphere, photochemical production in the lower troposphere, and destruction at the surface all play important roles.

**Troposphere–Stratosphere Exchange**

As indicated in the previous section, exchange between the stratosphere and troposphere is an important aspect of transport. The stratosphere and troposphere

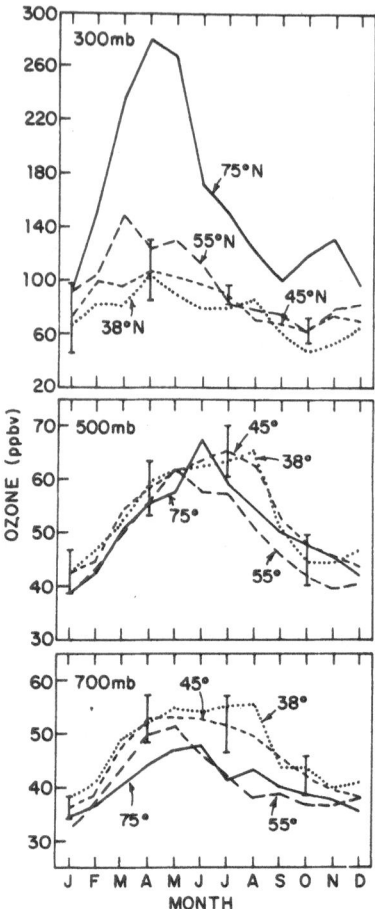

**Fig. 11.** The seasonal distributions of tropospheric ozone at several latitudes and heights. From [19]

are coupled not only through the exchange of trace chemical species, but also through exchange of momentum and energy. Such exchange is critical to understanding of the budget of ozone in both the stratosphere and the troposphere. It is also an important element in the overall dynamics of climate. The rate of movement across the tropopause of certain species that absorb and emit infrared radiation determines their global lifetimes, and hence their atmospheric concentrations. This in turn determines their contributions to the greenhouse effect. Thus, troposphere–stratosphere exchange must be regarded as one of the key problems in climate dynamics as well as an important issue for transport modeling.

In the discussion of cross tropopause exchange, it is useful to focus on two types of tracers. The first type are tracers with sources in the troposphere that are transported into the stratosphere where they are destroyed either by various photochemical processes or by radioactive decay (e.g., radon, nitrous oxide, methane, water vapor, and the various chlorofluoromethanes). The second type are tracers with sources in the stratosphere that are transported into the troposphere where they are removed by a number of processes (e.g., ozone, nitric acid, and stratospheric aerosols). Cross tropopause transport of tracers does not occur uniformly by slow diffusive processes. Rather the transport is concentrated in association with special meteorological conditions. Most of the upward transport appears to take place in the tropics, while most of the downward transport occurs in extratropical latitudes.

Although transport is important in all parts of the atmosphere, the vertical flux of tracers across the tropopause and through the lowest few kilometers of the stratosphere is especially significant since it determines the rate of flow between source and sink regions for a number of important molecules. The slowness of the transport in this region is responsible for the observed long (several year) residence time for radioactive bomb debris in the stratosphere (see, Telegadas and List, [20]). It is also the reason that, according to model predictions, decades are required for the ozone layer to adjust to a steady state perturbed balance for a specified steady state release of chlorofluoromethanes.

**Transport Across the Tropical Tropopause**

The simplest model for stratosphere–troposphere exchange consists of bulk advection by the diabatic circulation (Hadley cell), with uniform rising across the tropical tropopause, poleward drift in each hemisphere, and sinking at mid and high latitudes. This sort of circulation is the basis of the transport scheme proposed by Brewer [21]. He argued that the upward moving air must pass through the "cold trap" of the high cold tropical tropopause in order that sufficient water vapor be condensed out to account for the extreme observed aridity of the stratosphere (water vapor mixing ratios of order 3 ppm). Dobson [22] noted that the poleward and downward high latitude branch of this meridional circulation was consistent with the observed distribution of ozone in the lower stratosphere; ozone is a maximum at high latitudes, while the source of ozone is at equatorial latitudes. Although the so-called Brewer–Dobson cell represents a useful conceptual model of stratosphere–troposphere exchange, and transport within the lower stratosphere, it does not provide a physical model for the actual exchange process.

It is true that the overall circulation implied by the Brewer–Dobson model is consistent with the diabatic circulation discussed in previous section. That is, the upward motion at low latitudes is consistent with net diabatic heating, and the downward motion at high latitudes is consistent with diabatic cooling. The freeze drying of air due to upward passage through the tropical tropopause no doubt is the cause of the extremely low water vapor mixing ratios in the stratosphere. But recent observational studies have revealed a number of difficulties with the simple Brewer model. Dehydration of air to the mixing ratios observed would require that air entering the stratosphere experience temperatures of about 190 K in passing through the tropopause at the 10 kPa level (about 16 km), and that the ice crystals condensed out not be carried into the stratosphere, so that after passage into the stratosphere the resultant mixing ratios would be those corresponding to ice saturation at the tropopause.

Climatology, however, indicates that on average the tropical tropopause is not nearly cold enough to freeze dry air passing through it to mixing ratios of less than 4 ppm. In addition, a slow uniform ascent would imply existence of a thick layer of cirrus cloud near the tropical tropopause. Such is not observed. On the contrary, the clouds in the equatorial zone are primarily of a convective nature. This convection is not distributed evenly over the tropics, but is concentrated in a narrow intertropical convergence zone (ITCZ), which is generally found a few degrees from the equator in the Northern Hemisphere. As discussed in Houze [23], for example, nearly all of the upward transport in the tropical troposphere occurs in the updrafts associated with cumulonimbus convective clouds, and their associated mesoscale anvils. Such "hot tower" cloud systems may overshoot their levels of neutral buoyancy and penetrate into the lower stratosphere carrying water and other tracers.

Thus, it appears that cumulonimbus clouds might be the entities responsible for most of the vertical transport of trace species across the tropical tropopause. However, this process does not by itself explain some important observations. In Fig. 12 we show the range of water vapor and temperature profiles measured around the tropopause level by the NASA ER-2 research aircraft in Panama (8 °N) during September, 1980. The minimum in water vapor mixing ratio occurs at about 19 km. This minimum is well below the saturation mixing ratio at the tropopause level; it also occurs well above the tropopause, and much higher than the penetration depth for cumulonimbus cloud towers. Thus, there seems to be little possibility that the stratospheric dehydration observed in the Panama region could be due to local processes. Long range horizontal transport seems to be required to explain the observations.

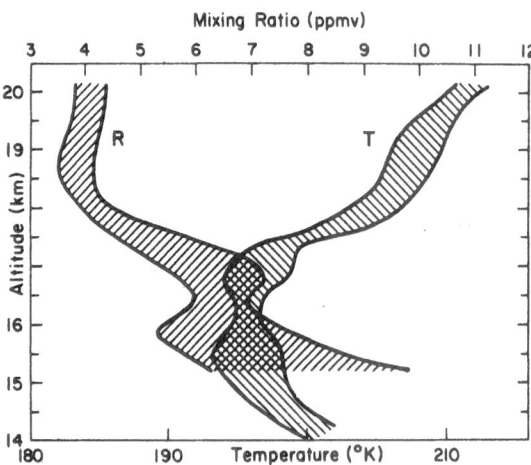

**Fig. 12.** The range of temperature (T) and water vapor mixing ratio (R) measured by the NASA U-2 aircraft in Panama during September 1980. From [24]

Newell and Gould-Stewart [25] suggested that the stratospheric water vapor observations are consistent with a model, which they call the stratospheric fountain, in which most of the flux of mass from the troposphere into the stratosphere is concentrated in a few areas of the tropics where the tropause is exceptionally high and tropopause temperatures are exceptionally cold. These areas are the Indian monsoon region during Northern hemisphere summer and the Indonesian "maritime continent" during Northern Hemisphere winter. Analysis of meteorological temperature soundings for a large network of tropical stations during the global weather experiment in 1979 [26] generally supported this notion, although during the Northern winter season tropopause temperatures sufficiently cold to account for the dryness of the lower stratosphere occurred over a wider range of longitudes than suggested by Newell and Gould-Stewart.

The stratospheric fountain hypothesis provides a possible explanation for the water vapor profile measurements over Panama. It is not necessary that the 4 ppm mixing ratios observed at 19 km there be produced by freeze drying during passage through the local tropopause. Rather, the dry air observed during the summer in Panama at 19 km may have entered the stratosphere over Indonesia during the Northern winter (where there is the potential to freeze dry air parcels to less than 3 ppm and through slow upward motion, longitudinal drift, and mixing with moister air might appear as a dry layer well above the tropopause in Panama.

Even if this conceptual model is correct, however, it is still necessary to explain the physical process by which the ice crystals that are formed during the freeze dry process are left in the troposphere while the air moves into the stratosphere. Overshooting cumulonimbus convective cells would be expected to carry their condensate into the stratosphere, where evaporation would occur. Thus, the immediate effect of the convective overshooting should be to moisten the stratosphere.

However the sinking due to negative buoyancy that follows convective overshooting produces massive cirrus anvil clouds in the upper troposphere and lower stratosphere. The anvils, which are much larger in horizontal extent than occur in midlatitude thunderstorms [23, 27, 28] may last for 5 to 10 hours or more. Because there is rapid radiative cooling to space from the top surface of the anvil, and radiative heating due to exchange of radiation with lower atmosphere layer at the bottom surface of the cloud, a moist adiabatic lapse rate is quickly established in the anvil, leading to small scale turbulent overturning.

Danielsen [27] proposed that the air passing into the stratosphere in association with tropical convection is processed by these cirrus anvil clouds. In his model, which has been partly confirmed by observations during the Stratosphere–Troposphere Exchange Project (STEP) based in Darwin Australia in early 1987 (Danielsen, personal communication), radiative cooling from the anvil tops tends to destabilize the anvil cloud to produce an upward flux of heat and vapor. The latter causes the ice crystals to grow and fall out so that at the end of the life cycle of the anvil the remaining air is dehydrated and has potential temperatures characteristic of the stratosphere.

### Transport Across the Extratropical Tropopause

In the Brewer–Dobson circulation model the transfer of mass from the stratosphere to the troposphere takes place through slow large scale subsidence in the extratropical region. Although the diabatic circulation is indeed downward in this region (see Fig. 7) the major exchange does not occur through slowly varying mean motions, but occurs in sporadic events associated with the development of extratropical storms. This cyclogenisis occurs preferentially in the region of strong meridional temperature gradients associated with subtropical and polar jetstreams.

An example of the potential temperature distribution associated with such a jet is shown in Fig. 13. The isentropic surfaces slope strongly so that air parcels sliding along the isentropes can move adiabatically from the lower stratosphere deep into the troposphere. This process, in which the material tropopause is advected by motions along the sloping isentropes so that it intrudes deeply into the troposphere to form a tropopause fold, provides a very efficient mechanism for irreversible mixing of stratospheric air into the troposphere. The stratospheric air that slides downward into the troposphere along the fold, as shown in Fig. 14, is gradually diluted as the material in the fold undergoes a stretching deformation until the cross fold scale may be as small as 1 km, and small scale mixing processes produce irreversible mixing. Evidence for this is clear in the gradual reduction of ozone mixing ratio shown for air moving downwards into the fold depicted in Fig. 14.

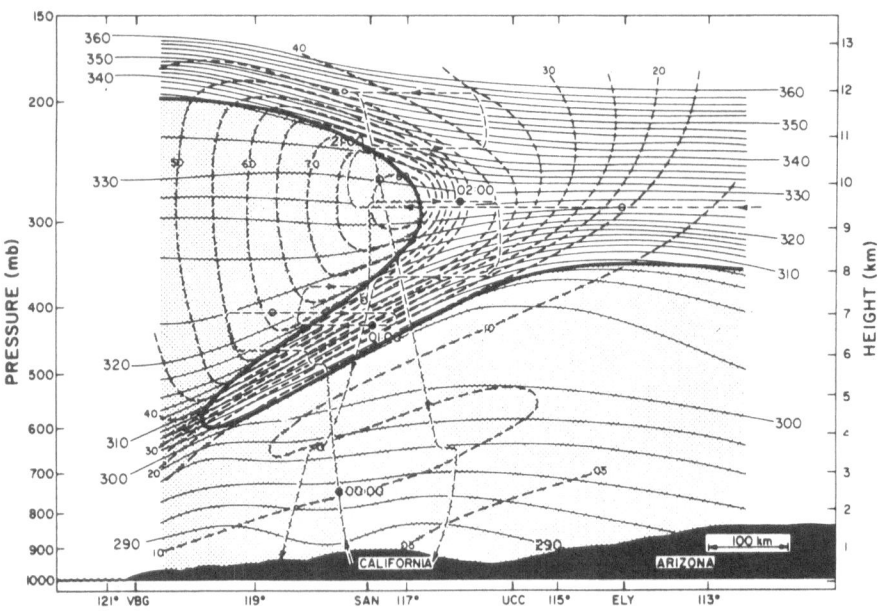

**Fig. 13.** Cross-section of a tropopause fold event. Region of tropospheric air is stipled; potential temperature (thin solid lines) wind speed (m s$^{-1}$, dashed lines); tropopause (heavy solid line) are all depicted. From [29]

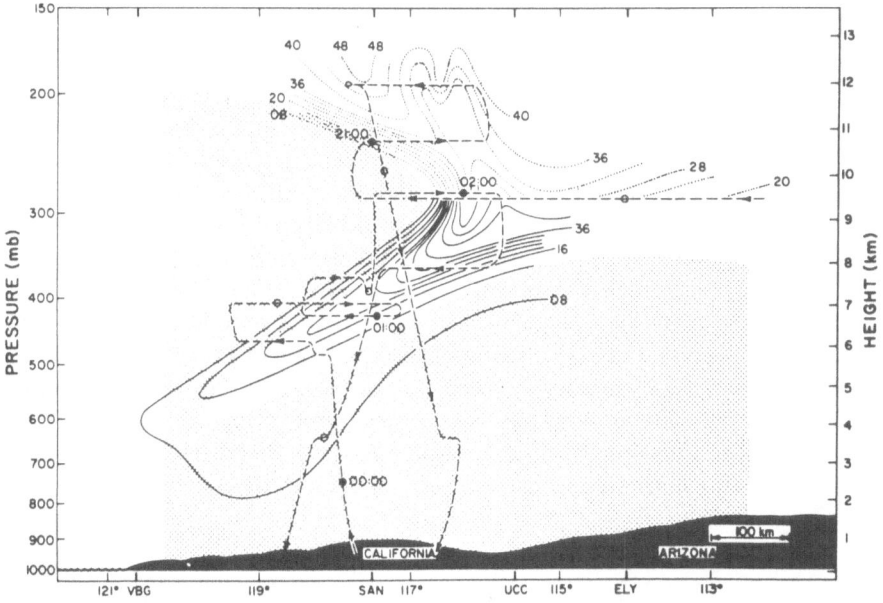

**Fig. 14.** Ozone mixing ratio in parts per hundred million by volume for tropopause folding event shown in Fig. 13. Dashed lines show flight tracks of research aircraft. From [29]

Although a number of estimates have been made of the total annual exchange of mass between the stratosphere and troposphere, such estimates provide little insight into the net rate of irreversible transport of trace molecules across the tropopause. Such rates depend on the distribution of sources and sinks for a given molecule. For most chemical tracers it is not simply the rate of transfer across the tropopause that is important, but the rate at which the molecule is transferred from its source region to its sink region. Thus, for molecules that are photochemically destroyed in the middle stratosphere any exchange that involves only the lowest part of the stratosphere is not important. The most useful measure of the magnitude of global exchange rates is perhaps given by the strength of the diabatic circulation in the region of the tropopause. Unfortunately, the net diabatic heating in this region is a small residual difference between large heating and cooling terms, and is thus difficult to evaluate accurately [30].

## Transport Within the Stratosphere

As pointed out above, the earliest studies of transport in the stratosphere were stimulated by efforts to understand the budgets of water vapor and ozone. By the 1960s, however, the focus of global transport studies shifted towards attempts to observe and account for the evolution of radioactive debris released into the stratosphere in the course of atmospheric nuclear testing. Mahlman et al. [11] give an interesting survey of the history of efforts to understand transport in the lower stratosphere. Their analysis clearly distinguishes between the roles of eddy and mean flow processes in contributing to net transport.

Because of its importance for understanding the ozone layer, tracer transport within the stratosphere has been much more extensively studied in recent years than has tropospheric transport. Unlike the case for the troposhere, where global distributions of trace chemicals are generally only poorly known, satellite observations have provided a substantial amount of global data for several stratospheric tracers. Ozone and stratospheric aerosols have been observed by several satellites, while other species such as $CH_4$, $N_2O$, and $HNO_3$ have been observed by instruments on the Nimbus 7 satellite. These observations show without doubt that for long lived tracers, the major variations in the stratosphere are in latitude and height; longitudinal variability is rather small except in the winter hemisphere where rapid meridional transport associated with transient planetary wave disturbances can occasionally produce large variances. Thus, the primary focus of stratospheric transport studies has been to understand tracer distributions in the meridional plane.

## Long-lived Tracers

Much can be learned by considering the climatological distribution of quasi-conservative tracers in the stratosphere. The role of transport in determining tracer climatology depends on the nature and distribution of tracer sources and sinks, and on the relative magnitudes of the timescales for dynamical and chemical processes. It is useful to distinguish here between tracers whose sources are mainly in the troposphere and those whose sources are mainly in the stratosphere. The former class (e.g., nitrous oxide and methane) are transported into the stratosphere where they are destroyed through photolysis or oxidation; the latter class (e.g., ozone and cosmogenic radionucleides) are transported into the troposphere where they are removed by several processes.

The distribution of these substances in the stratosphere is dependent on the competition between dynamical and chemical processes. This competition can be approximately measured by comparing the characteristic dynamical and chemical timescales. By *chemical timescale* is meant the average time for replacement or removal of a tracer due to local sources and sinks. by *dynamical timescale* is meant the average time for advective and diffusive processes to transport the tracer from equator to pole or across a scale height in the vertical.

If the chemical timescales for both sources and sinks are much shorter than the dynamical timescale, the tracer will be in local photochemical equilibrium and transport does not directly influence its distribution. Transport can, however, play an important indirect role by partly determining the concentrations of other species that participate in the photochemical production or loss of the tracer in question. If the chemical timescale is much longer than the dynamical timescale, the tracer will be passively advected by the flow field. In the absence of localized sources and sinks such a tracer will eventually become well mixed due to the dispersive effects of transport.

When the chemical and dynamical timescales are comparable the observed species concentration depends on the net effects of chemical sources and sinks and of transport. In many cases of interest for the stratosphere the ratio of chemical and dynamical timescales changes drastically with altitude in the stratosphere. Several examples are shown in Fig. 15. Such tracers, with sources in the troposphere have vertical profiles with decreasing mixing ratios with height (Fig. 16). At any level the mean concentration results from a balance between the chemical destruction and the net flux convergence. Although Fig. 16 suggests that it is a vertical flux convergence that balances chemical loss (since the mixing ratio decreases with height) the actual situation at any location is more complicated due to the strong meridional dependence of the mixing ratio in the stratosphere for these species. In all cases these tracers have very long lifetimes in the troposphere and the lower stratosphere and will be referred to here as long-lived tracers since the bulk of the tracer mass is contained in the troposphere and lower stratosphere.

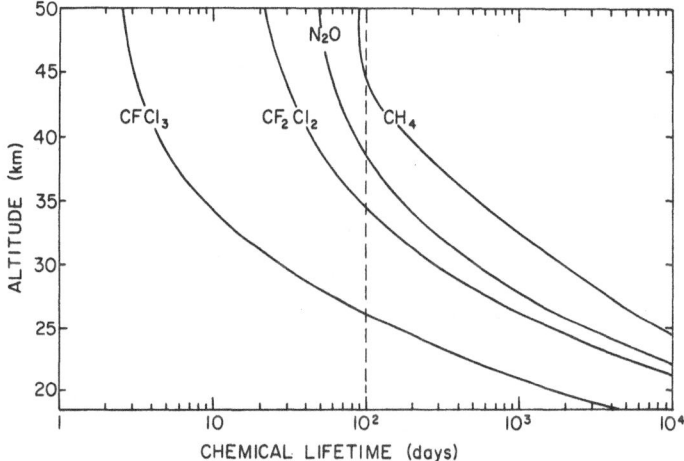

**Fig. 15.** Altitude dependence in the stratosphere of chemical lifetimes (days) for several tropospheric source gases. Dashed line shows typical transport timescale in the stratosphere. From [31]

**Transport in the Meridional Plane**

The role of transport in determining the distribution of long-lived tracers can best be appreciated by considering the tracer distribution in the meridional plane. For methane and nitrous oxide monthly mean distributions have been observed by the Stratospheric and Mesosphereic Sounder (SAMS) instrument on the Nimbus 7 satellite. Although there are important seasonal variations, both these species tend

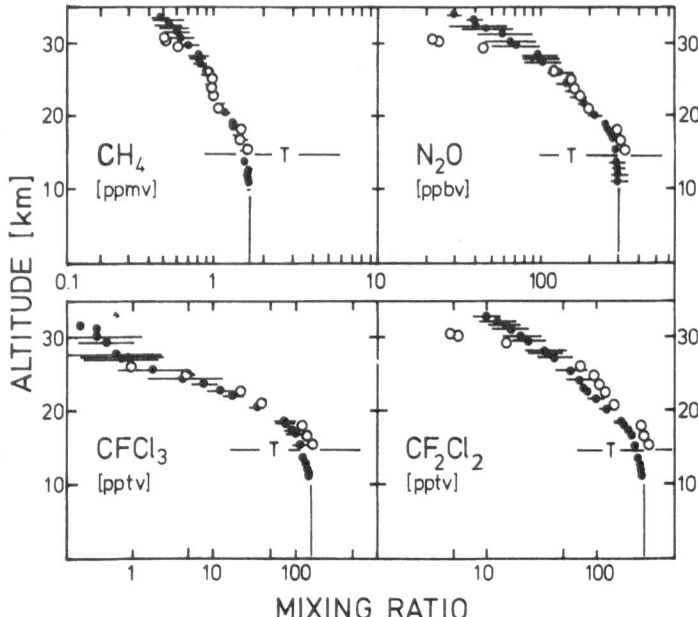

**Fig. 16.** Midlatitude vertical profiles of mixing ratio for several tropospheric source gases from balloon observations over Southern France. From [32]

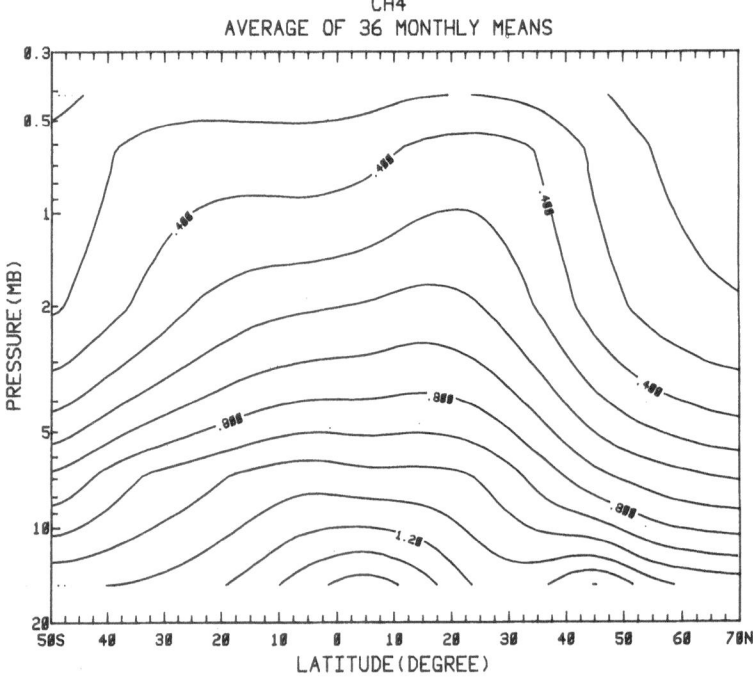

**Fig. 17.** Three year average of zonally averaged $CH_4$ mixing ratio from the SAMS experiment on the Nimbus 7 satellite. Note that the data extends from 50 °S to 70 °N latitude

to have constant mixing ratio surfaces that are bulged upward in the tropics, and slope downward toward the poles in both hemispheres (Fig. 17). Since isentropic surfaces in the stratosphere tend to have small slopes, it is clear that the mixing ratios for $CH_4$ and $N_2O$ generally decrease with distance from the equator on isentropic surfaces.

As was indicated in an earlier section, no matter how the flow is formally partitioned between "mean" and "eddy" parts, the circulation in the meridional plane should be regarded as an eddy driven circulation. In the absence of departures from zonal symmetry, the stratosphere would be near radiative equilibrium, and the diabatic circulation in the meridional plane would be very weak. It cannot be stressed too much that it is eddy motions that, through their poleward heat transport, drive the stratosphere away from radiative balance, and hence produce the diabatic mean circulation required to balance the resulting radiative heating and cooling. Rather than focusing on the issue of mean versus eddy transport, it is probably more sensible to distinguish between advective and diffusive transport while recognizing that eddies may contribute to both.

The mean distributions of long-lived tracers such as $N_2O$, $CH_4$, $CFCl_3$, and $CF_2Cl_2$ are qualitatively those expected for tracers advected by the diabatic circulation, since the diabatic circulation would be expected to push the constant mixing ratio surfaces upward in low latitudes and downward at high latitudes. However, the observed slopes of the mixing ratio surfaces for these tracers are substantially less than would be produced by the diabatic circulation alone. In addition to the advection by the diabatic circulation, long lived tracers are influenced by quasi-isentropic mixing due to large scale eddy motions. The equilibrium tracer mixing ratio slopes are produced by a competition between slope steepening effects of the diabatic circulation, and slope flattening effects of the quasi-isentropic mixing by eddies. This competition among the advective and diffusive transports is illustrated in Fig. 18.

Except in its lowest layers, the stratosphere does not have the energetic synoptic scale eddies that account for much of the dispersive tracer transport in the troposphere. Dispersion in the stratosphere thus tends to be rather slow except in the winter hemisphere where planetary scale wave disturbances may produce large meridional excursions of the motion. These waves tend to separate the winter hemisphere into two regions: a polar vortex core, which is protected from lateral mixing, and a broad mid and low latitude region where the planetary waves may, through a process referred to as "planetary wavebreaking" (McIntyre and Palmer, [33]) cause rapid irreversible mixing of the tracers on isentropic surfaces.

The tendency for meteorological fields to generate such irreversible deformation is illustrated in Fig. 5, which shows the tendency of flow deformation to rapidly string a tracer blob out into long thin laminae that can be rapidly mixed by small scale turbulence. Such wavebreaking on a hemispheric scale is suggested by the distributions of $CH_4$ on the 3 mb pressure surface in the stratosphere shown for December 5, 1981 in Fig. 19. The tendency for the polar minimum to be drawn out westward and equatorward in a "tongue", and for the midlatitude maximum to penetrate poleward and eastward are characteristic of the "wrapping up" of tracer by planetary wave breaking, and clearly lead to a net poleward flux of tracer.

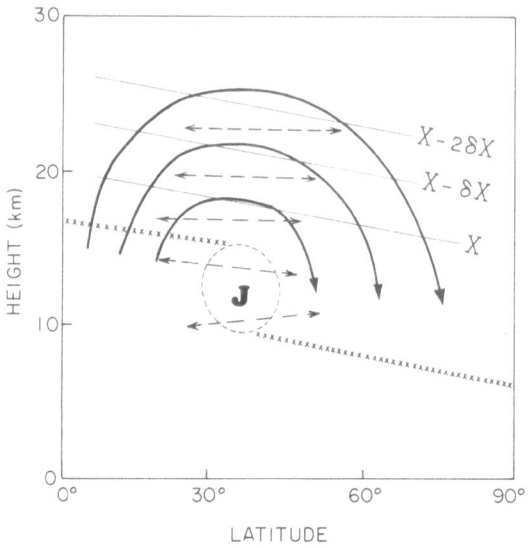

**Fig. 18.** Schematic view of the competition between advective and diffusive transport in the strato-sphere. Heavy lines show the mean meridional mass circulation. Dashed lines indicate quasi-isentropic mixing by large-scale eddies. The mean tropopause is marked by crosses, and J indicates the mean jetstream core. Light lines labeled with mixing-ratio values (X) show mean slope of a long-live vertically stratified tracer. From [8]

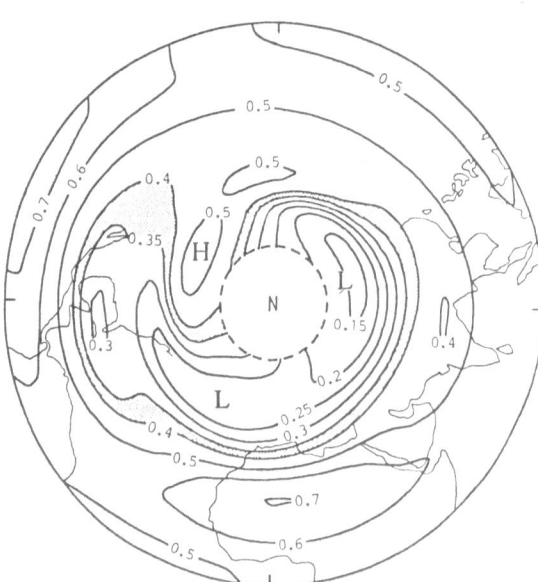

**Fig. 19.** Polar stereographic chart of the $CH_4$ mixing ratio contours (ppmv) at the 3 mb pressure level on December 6, 1981 as observed by SAMS. Note the tongue of low mixing ratio air that has been drawn out of the polar region and extends across the North American continent. From [34]

### Distribution and Transport of Ozone

Atmospheric ozone has been observed remotely using ground based, and satellite instruments. It has also been observed directly using balloon and aircraft platforms. Most of the remote observations provide only the total column abundance, rather than the vertical distribution. Such observations are not as useful for transport studies as are observations of the local mixing ratio. Nevertheless, the latitudinal and seasonal variations of total ozone provide considerable indirect information on transport.

Total ozone is generally expressed in terms of the equivalent thickness of the ozone layer at standard temperature and pressure (STP; 0 °C, 1013.25 mb). The units used for this purpose are called Dobson units (DU) where $1\,DU = 10^{-5}\,m$ (STP). The global mean total ozone is about 300 DU. A four year average of the global distribution of total ozone as observed by the Nimbus 7 total Ozone Mapping Spectrometer (TOMS) is shown in Fig. 20. Despite the fact that ozone is photochemically produced primarily in the middle and upper stratosphere at low latitudes, the column ozone has a broad equatorial minimum and maxima at high latitudes in both hemispheres. This indicates that the ozone distribution is strongly influenced by transport, and that transport must, on the average, carry ozone from low latitudes to high latitudes.

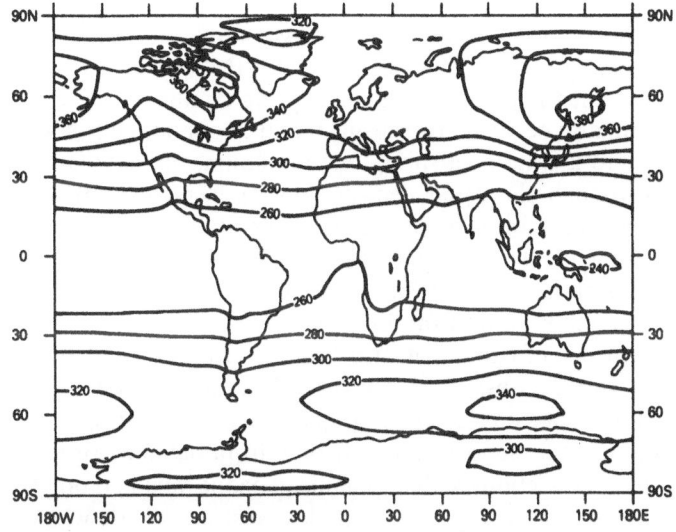

**Fig. 20.** Global distribution of total ozone (Dobson units) based on 4 years of observations with the TOMS instrument on Nimbus 7. From [35]

As is clear from Fig. 20 the time mean total ozone field is primarily dependent on latitude, not longitude. The longitudinal dependence is also small on the monthly time scale. Thus, in displaying the seasonal variability, it is sufficient to plot the longitudinally averaged total ozone in the time latitude domain (Fig. 21). The

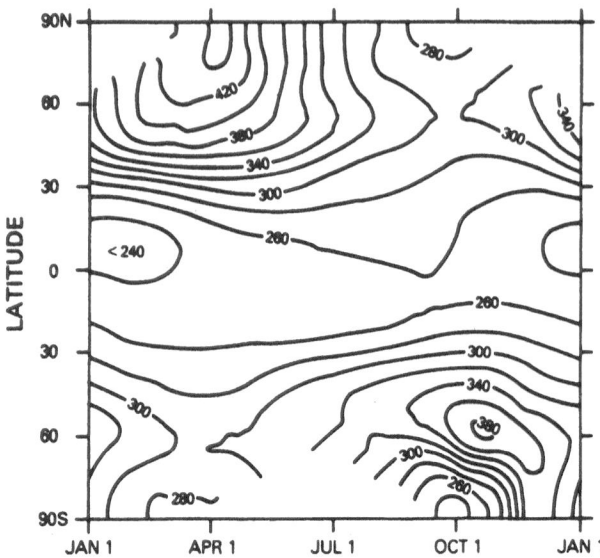

**Fig. 21.** Time-latitude section showing the seasonal variation of total ozone (Dobson units) based on TOMS data. Note the spring maximum near 90 °N and 60 °S and the October minimum at 90 °S (the ozone hole). From [35]

seasonal variability is dominated by a large amplitude annual cycle in the extratropical regions of both hemispheres. In the Northern Hemisphere the maximum occurs at the pole at about the spring equinox. The Southern Hemisphere maximum also occurs in the spring, but at about 60°S latitude. Poleward of that latitude there is a notable springtime minimum, the now famous Antarctic ozone hole. Although the rapid decrease during September that leads to the formation of the ozone hole is thought to be due mainly to chemical processes, it is plausible to assume [5] that transport plays a role in that the polar vortex forms an isolated material entity within which anomalous chemistry can occur without disruption by mixing in of air from lower latitudes.

The vertical profile of ozone has also been determined globally from satellite. However, most satellite observations provide little useful information below the level of maximum ozone concentration (especially in the troposphere), so that the satellite profiles must be supplemented by data from balloon borne ozonesondes. Latitude-height cross sections of the zonal mean ozone mixing ratio profiles for the four seasons are shown in Fig. 22. In the lower stratosphere, where the photochemical timescale is long enough so that transport dominates in the ozone budget, the mixing ratio surfaces have poleward-downward tilts similar to those of the long-lived tracers discussed in the previous section. Such distributions are clearly consistent with poleward and downward advection by the diabatic circulation. Hence, the global ozone distribution is strongly influenced by global transport processes that remove the ozone from its chemical source region above 30 km in the tropics, into the lower extratropical stratosphere where stratosphere–troposphere exchange processes can remove it into the troposphere.

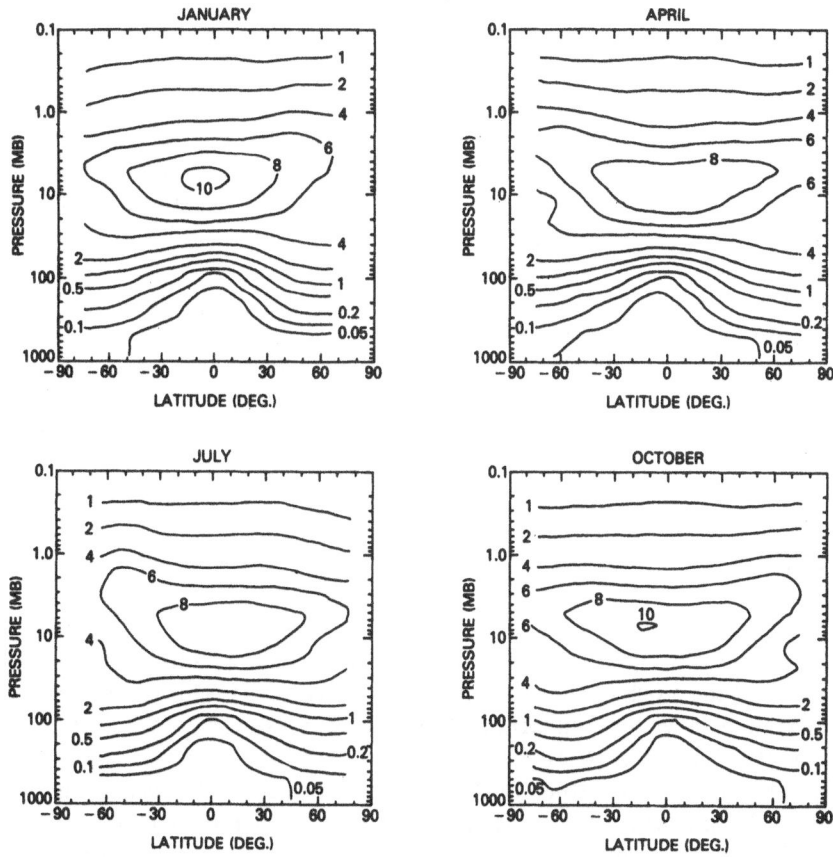

**Fig. 22.** Latitude-height sections of the ozone mixing ratio (ppmv) for January, April, July, and October 1979, as observed by the Nimbus 7 satellite. From [36]

## Transport Modeling

Transport models have become essential tools in the theoretical study of human influences on the atmosphere. Although models are now being used to investigate a number of problems of global tropospheric chemistry, it is fair to say that to date the largest efforts in global transport modeling have been devoted to the problem of human caused perturbations of the ozone layer. Since nearly all of the chemical species thought to be relevant for the ozone budget are directly or indirectly influenced by transport, and since species concentrations and chemical processes relevant to the ozone budget tend to have strong gradients in height, many of the studies of ozone layer perturbations have employed models that resolve only the vertical variability.

Such one-dimensional models can only parameterize transport processes in a very crude fashion, and of course provide no direct information on latitudinal variability. For this reason, and because photochemical processes tend to have strong latitudinal dependencies, the recent emphasis in ozone layer modeling has been on two-dimensional models that can simulate the distribution of ozone in the latitude-height (meridional) plane and can incorporate the seasonal cycle. However, even these models must parameterize the important effects of transport by zonally asymmetric eddies. None of the techniques yet derived to do this are completely satisfactory. Thus, attempts are now being made to use three-dimensional general circulation models, which explicitly resolve the major eddy fluxes as the basis for transport studies.

The three dimensional model approach is especially important for study of tropospheric transport where sources and sinks may be distributed very irregularly in longitude due to continent–ocean differences. However, this sort of modeling is still in its early stages. Many of the past studies of global tropospheric transport have been made on the basis of highly simplified "box" models whose nature is described in the next section.

## Box Models

The simplest transport model partitions the atmosphere into a small number of reservoirs, or boxes. When the inflow of a trace substance into a box equals the outflow rate, the stock of substance in the box is constant. This is referred to as a steady state balance, or equilibrium. The ratio of inflow (or outflow) rate to total tracer mass in a box is referred to as the turnover time (or residence time). Box models are popular in the study of global biogeochemical cycles such as the carbon cycle and the nitrogen cycle (see, for example, Siegenthaler and Oeschger, [37]). In such cases the atmosphere, the oceans, the solid earth, and the biosphere are each represented as boxes, which may contain various sources and sinks, and are connected through fluxes of material between them.

In considering the atmosphere in isolation there are two types of box models that are especially useful. In the first the separation is between Northern and Southern hemispheric boxes, while in the second it is between tropospheric and stratospheric boxes.

Harte [38] has provided a simple example of a two box model that can provide useful information on interhemispheric transport rates in the troposphere. In his model the budget of ethane is computed for two boxes, one representing the Northern hemisphere and the second the Southern hemisphere.

Ethane is emitted into the atmosphere almost entirely through escape during natural gas production. The concentration in a given hemisphere of the atmosphere is dependent on emission rates, destruction rates, and the fluxes between the hemispheres. The situation is shown in Fig. 23. For convenience the total mass M is expressed in moles, the emission rate E and interhemispheric fluxes F in units of moles/year, and the destruction rate is modeled as linearly proportional to the tracer mass in a given hemisphere, with proportionality coefficient $\alpha$. In all cases the subscripts N and S denote the Northern and Southern hemispheres, with the fluxes labelled according to the hemisphere from which the substance or flux originates.

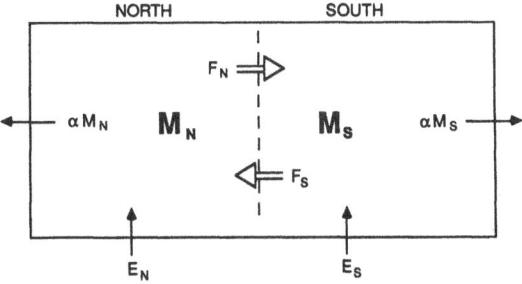

**Fig. 23.** Two box model for the ethane budget in the troposphere. Subscripts N and S denote Northern and Southern Hemispheres, respectively. See text for details

The budget equations for the masses $M_N$ and $M_S$ depicted in Fig. 23 are

$$\frac{dM_N}{dt} = E_N + F_S - F_N - \alpha M_N \tag{8a}$$

$$\frac{dM_S}{dt} = E_S + F_N - F_S - \alpha M_S. \tag{8b}$$

The cross equatorial flux is proportional to the tracer concentration divided by the turnover time. Thus, the fluxes out of the Northern and Southern Hemispheres, respectively, are $F_N = \beta M_N$, and $F_S = \beta M_S$ where $\beta$ equals the mass flux of air across the equator divided by the mass of air in a hemisphere. For steady state conditions the budgets are

$$E_N - \beta(M_N - M_S) - \alpha M_N = 0 \tag{9a}$$

$$E_S + \beta(M_N - M_S) - \alpha M_S = 0. \tag{9b}$$

Suppose for simplicity that $E_S = 0$. This is reasonable for ethane since nearly all of the natural gas production is in the Northern Hemisphere. Then if the coefficients $\beta$ and $\alpha$ are known (9) can be solved for the concentrations $M_N$ and $M_S$. Alternatively, if the concentrations in the two hemispheres are known from observations, then (9) can be solved for the turnover time:

$$\beta^{-1} = (M_N^2 - M_S^2)(E_N M_S)^{-1}. \tag{10}$$

According to [38], $E_N = 1.2 \times 10^{11}$ moles per year, and $M_N = 2M_S = 0.9 \times 10^{11}$ moles. Thus from (10) $\beta^{-1} = 1.125$ years. Hence, knowledge of only the emission rate and the mean concentration in each hemisphere is sufficient to determine the interhemispheric exchange time.

A second type of box model that can provide insights into global transport in the atmosphere is the two box model shown in Fig. 24. The lower (tropospheric) box occupies the pressure range from 100 to 10 kPa, while the upper (stratospheric) box occupies the range 10 to 0 kPa. This model is appropriate for species that are emitted into the troposphere and destroyed in the stratosphere. We let $M_T$ and $M_S$ denote the mass per unit area (kg m$^{-2}$) in the troposphere and stratosphere, and S,

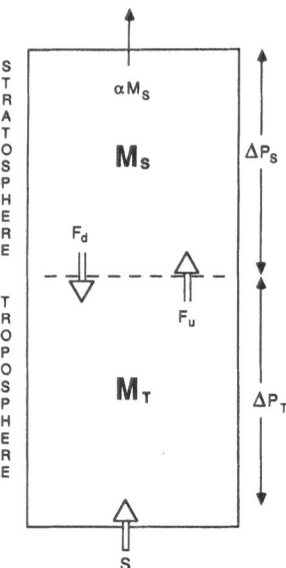

**Fig. 24.** Two box model for a tropospheric source gas (subscript T) that is destroyed in the stratosphere (subscript S). See text for details

$F_u$, and $F_d$ be the emission rate, and the upward and downward fluxes across the tropopause $(\text{kg m}^{-2}\text{s}^{-1})$, respectively. We also assume that the destruction is entirely due to photochemical processes that are limited to the stratosphere, and has a rate $\alpha$ proportional to the concentration. The continuity equations for the column masses in each box are then

$$\frac{dM_S}{dt} = F_u - F_d - \alpha M_S \tag{11a}$$

$$\frac{dM_T}{dt} = S - F_u + F_d. \tag{11b}$$

The fluxes are most conveniently expressed in the form $F_u = \chi_T W$, and $F_d = \chi_S W$, where W is the rate of air mass exchange across the tropopause $(\text{kg m}^{-2}\text{s}^{-1})$, and $\chi$ designates a mass mixing ratio. Noting that the mass per unit horizontal area in the troposphere and stratosphere can be expressed as $\Delta p_T/g$ and $\Delta p_S/g$, respectively, the tracer column mass can be expressed in terms of the tracer mixing ratio as follows:

$$M_T \simeq \chi_T(\Delta p_T g^{-1}); \quad M_S \simeq \chi_S(\Delta p_S g^{-1}). \tag{12}$$

Substituting from (12) into (11) the budget equations for tracer mixing ratio can be expressed in the form

$$\frac{d\chi_S}{dt} = -\beta_S(\chi_S - \chi_T) - \alpha\chi_S \tag{13a}$$

$$\frac{d\chi_T}{dt} = \frac{S}{(\Delta p_T g^{-1})} - \beta_T(\chi_T - \chi_S) \tag{13b}$$

where $\beta_S = W/(\Delta p_s/g)$ is the inverse of the turnover time for the stratosphere and $\beta_T = W/(\Delta p_T/g)$ is the inverse of the turnover time for the troposphere. (Note that the turnover time for the troposphere exceeds that for the stratosphere by the ratio of their masses, $\Delta p_T/\Delta p_s$). At steady state the mixing ratios in the stratosphere and the troposphere are given by

$$\chi_S = \frac{\alpha^{-1}S}{(\Delta p_s g^{-1})}; \quad \chi_T = \chi_S(1 + \alpha \beta_S^{-1}).$$

Thus the turnover time for the troposphere can be estimated from a knowledge of $\chi_S$, $\chi_T$, and $\alpha$ for a tracer in steady state balance that has no sinks in the troposphere. $N_2O$ has an atmospheric burden of order $1.5 \times 10^9$ tons and an annual source of about $15 \times 10^6$ tons per year [3], which implies approximately a 100 year turnover time for the troposphere, and a 10 year turnover time for the stratosphere.

## One-dimensional Models

Although the variation of ozone and its photochemistry with latitude is quite large, many of the most important aspects of the stratospheric ozone budget can be studied in a model in which only the altitude profile is explicitly retained. A relatively complete theoretical treatment of the photochemistry of the stratosphere requires a model in which about 100 chemical equations are included and the effects of transport are calculated for a dozen or so chemical species. Thus, the one-dimensional model, because of its comparitive simplicity, is a very popular tool for study of potential human perturbations of the ozone layer.

A one-dimensional tracer continuity equation can be derived from the two or three dimensional tracer continuity equation by averaging globally on constant altitude, pressure, or potential temperature surfaces. In such a model, of course, the concentrations of the various tracers must be regarded as global averages, and the model should be compared to globally averaged observations. In practice, global distributions are known for only a small fraction of the chemical species important in the ozone problem, and the one-dimensional model profiles are usually assumed to be characteristic of profiles at 30° or 45° latitude, and compared with observed profiles at those latitudes.

In a typical one-dimensional model the mass mixing ratio of a given species $\chi$ is given by the continuity equation

$$\rho \frac{\partial \chi}{\partial t} = \frac{\partial}{\partial z} \left[ \rho K_{zz} \frac{\partial}{\partial z}(\chi) \right] + P - L \tag{14}$$

where P and L are the photochemical production and loss rates ($kg\,m^{-3}\,s^{-1}$), $\rho$ is the air density and $K_{zz}$ is the vertical diffusion coefficient ($m^2\,s^{-1}$).

Generally the diffusion coefficient has been determined empirically by fitting the model to the observed distribution of a long-lived tracer such as $CH_4$ at 30°N. In such empirical formulations the same diffusion coefficient is then employed for all other tracers included in the model.

This approach would be reasonable if the vertical transport being parameterized by such a model was due to small scale turbulence that locally mixed columns in the vertical. In such a case the tracer budget for a given column would not depend on the concentration of tracer at other latitudes and longitudes. However, in reality the major vertical transport in the stratosphere is due to the global scale diabatic circulation. As shown in Fig. 18, for tracers with tropospheric sources the tracer mixing ratio is relatively large where the diabatic motion is upward (low latitudes) and is relatively small where the diabatic motion is downward (high latitudes). Thus, the vertical motion and the tracer mixing ratio are positively correlated so that there will be a net upward tracer flux when averaged over the whole globe at a given level. The strength of this correlation depends on the steepness of the mixing ratio surface slope, which varies from one molecule to another. Therefore, the appropriate vertical diffusion coefficient also varies for different species. Mahlman et al. [39] and Holton [31] have shown that when these factors are considered the appropriate value of diffusion coefficient is given by

$$K_{zz} = \langle W^{*2} \rangle \langle \tau_d^{-1} + \tau_c^{-1} \rangle^{-1} \tag{15}$$

where $W^*$ is the diabatic vertical velocity, $\tau_d$ is the timescale for hemispheric scale mixing, $\tau_c$ is the chemical timescale for the species to be modeled, and $\langle \ \rangle$ denotes a global average at constant height. A sample calculation of vertical diffusion coefficients for the tracers of Fig. 16 is shown in Fig. 25.

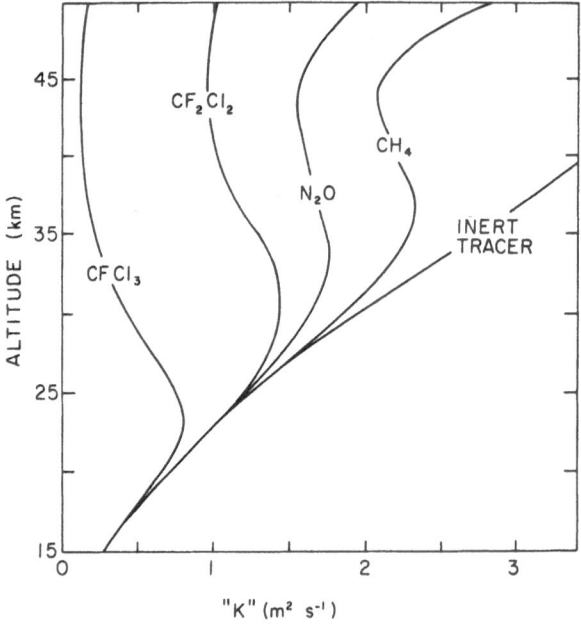

**Fig. 25.** Vertical eddy-diffusion coefficients for the trace gases whose lifetimes were shown in Fig. 15, based on the one dimensional model discussed in the text. From [31]

### Two-dimensional Models

Among the most popular models for study of global transport are two-dimensional models, which resolve tracer variability in latitude and height, but are averaged over longitude. Longitudinal averaging has proved to be very valuable in modeling the stratospheric ozone layer, and has also been useful in modeling the seasonal and longer term variability of long-lived tropospheric tracers, particularly those species with seasonal cycles that are strongly latitudinally dependent (e.g., $CO_2$ and $CH_4$).

As indicated in an earlier section there are several mathematical formulations for two-dimensional models, which differ according to the type of longitudinal averaging process that is employed. To date, the two-dimensional models that have been used for study of global *tropospheric* tracer problems have nearly all employed the conventional Eulerian mean formulation given in equation (4) with the eddy flux terms parameterized in terms of the zonal mean fields in the form of diffusion in the latitude-height plane. Examples are the models described in [40] and [41]. Such models have been used, for example, in studies of the global transport of carbon dioxide [42], and in studies of trends in the tropospheric concentrations of such trace gases as $O_3$, OH, CO, $CH_4$, and $NO_x$[43].

In this type of model the transport is split into two processes: advection by the Eulerian meridional circulation (Term A in (4)) and the eddy flux divergence (Term B in (4)). The mean meridional circulation in tropospheric models is generally obtained from climatological data for either monthly or seasonally averaged conditions. The eddy flux term is parameterized in terms of diffusion in the meridional plane:

$$\begin{bmatrix} \overline{v'\chi'} \\ \overline{w'\chi'} \end{bmatrix} = - \begin{bmatrix} K_{yy} & K_{yz} \\ K_{zy} & K_{zz} \end{bmatrix} \begin{bmatrix} \partial\bar{\chi}/\partial y \\ \partial\bar{\chi}/\partial z \end{bmatrix} \tag{16}$$

where the coefficients that form the elements of the diffusion tensor are empirically derived so that the model reproduces the distribution of some standard tracer. Most models, following Reed and German [44], have assumed that the diffusion tensor is symmetric so that $K_{yz} = K_{zy}$. However, this assumption, which is based on the mixing length theory of turbulent diffusion, is not appropriate for transport due to organized wave motions.

The major difficulty with this approach, as was pointed out earlier, is that the traditional Eulerian averaging procedure does a poor job of separating bulk advective and diffusive transport processes. A large fraction of the transport due to the eddy flux terms in this formulation is really of advective, rather than diffusive, nature. Furthermore, there is a large degree of cancellation that often occurs between the mean meridional circulation transport, and the transport represented by the eddy flux terms. Thus, the conventional Eulerian formulation is not a very efficient method of modeling global transport in the meridional plane. Nevertheless, models based on this scheme have produced useful information on a number of tracers.

The earliest two-dimensional models designed to study the stratospheric ozone layer were also based on the conventional Eulerian formulation (see [3], for a

comprehensive review), and some of these models are still used for ozone perturbation studies. However, most of the two-dimensional models currently used for
study of the ozone layer are based on the isentropic model (equation 6) or
transformed Eulerian models that split the advective and diffusive parts of the
transport in a manner that is very similar to that of the isentropic model.

In most such two-dimensional models the diabatic circulation (i.e., the mean
meridional circulation in the isentropic coordinates) is specified based on theoretical calculations of the radiative heating rates in the stratosphere for the various
seasons. An example is the model of Ko et al. [45], which has the meridional
streamline pattern at solstice and equinox shown in Fig. 26. Note that at the
equinox the two cell pattern, with rising in the tropics and sinking at high latitudes,
extends throughout the model domain; but at the solstice this pattern is replaced
above 25 km by a single circulation cell in which there is radiative heating and rising
motion in the summer hemisphere, a cross equatorial drift, and radiative cooling
and sinking motion in the winter hemisphere, Ko et al., argued that with this
diabatic model formulation the transport of tracers in the stratosphere could be
modeled accurately without including vertical diffusion, and with only a small
amount of diffusion along the isentropes. The global distributions of $N_2O$ predicted
by this model for solstice and equinox are shown in Fig. 27. In this study the mixing
ratio of $N_2O$ was specified at the lower boundary, so it is not surprising that the

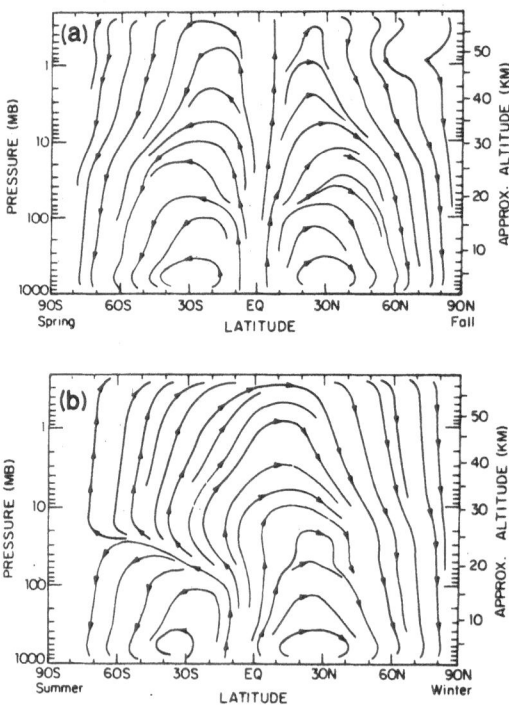

**Fig. 26.** Streamlines of a theoretical estimate of the diabatic circulation for (a) equinox and (b) solstice.
From [45]

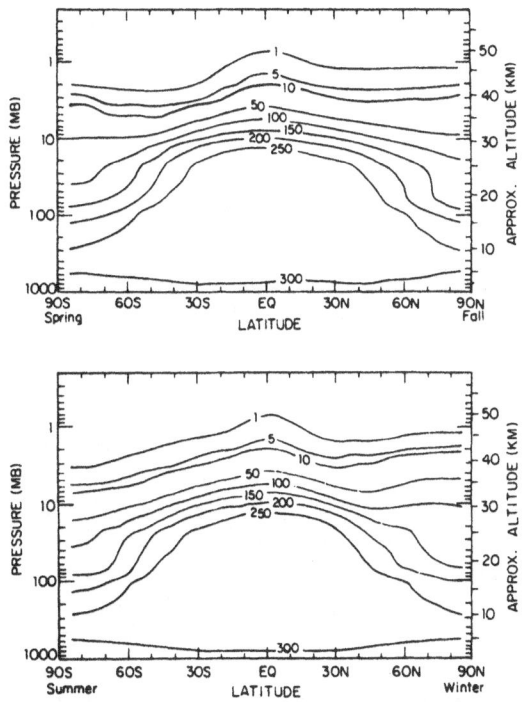

**Fig. 27.** Latitude-height section of $N_2O$ mixing ratio (ppbv) computed with a two dimensional model using the diabatic circulation of Fig. 26 and a horizontal diffusion coefficient of $10^5 \, m^2 \, s^{-1}$. From [45]

overall values are correct in the troposphere. More interesting is the distribution in the stratosphere where the model is able to reproduce the observed upward bulge of the mixing ratio surfaces in the tropics and the downward slope toward high latitude. These results tend to support the authors' suggestion that global scale transport in the stratosphere is primarily of an advective nature.

The model of Ko et al., and most other two-dimensional transport models use specified mean circulation fields, and can provide little information on the feedbacks between transport and dynamical processes. In other models the zonally averaged momentum and thermodynamic energy equations are solved (in addition to the tracer continuity equation) so that the mean meridional circulation is self consistently calculated, rather than being specified externally. Models in which the zonal mean dynamical equations are solved must, of course, employ parameterizations of the eddy forcing in the zonal mean momentum and thermodynamic equations. In one such model that has been used extensively for stratospheric chemical studies [46, 47] the eddy forcing in the momentum equation is a simple linear relaxation of the mean wind towards zero. Since the diabatic heating rate, which determines the mean meridional circulation, depends on the ozone distribution, which is influenced by transport, this type of model is able to incorporate at least some aspects of the interactions among radiation, dynamics and chemistry in the stratosphere. However, there is no obvious way to evaluate the possible role of such feedbacks in changing the role of eddies in transport. In fact, the parameterization of eddy fluxes remains the weakest aspect of two-dimensional modeling.

### Three-dimensional Models

Three-dimensional transport models are able to explicitly calculate transport effects due eddies. Thus, three-dimensional models that have sufficient resolution to accurately represent important meteorological features, such as transient cyclones and anticyclones, potentially should be able to provide a much better representation of global transport processes than can be done in two-dimensional models since the unresolved motions whose effects must be parameterized are much less energetic than is the case for two-dimensional models. Such models also have the advantage of being able to resolve the longitudinal structure of the tracer distribution, which can be important in the study of tropospheric tracers whose sources may have strong geographic variability. Three-dimensional tracer models are, however, very demanding of computer resources, and thus developments of such models have proceeded rather slowly. However, this is currently an active area of research, and rapid progress should occur in the next several years.

As in the case of the two-dimensional model, a first requirement for a three-dimensional transport model is a set of velocity vectors (i.e., winds) that can be used to transport the relevant tracers. Such a set could be obtained by using observed meteorological fields. However, observations are rather sparse over the oceans and in the Southern Hemisphere and are made at only 12 or 24 hourly intervals so that it is difficult to obtain a consistent set of time dependent observed winds with sufficient spatial and temporal resolution for accurate modeling of global transport processes.

Because of the difficulty of using observed velocities, most of the three-dimensional modeling efforts to date have utilized velocity fields obtained from the output of three-dimensional general circulation models (GCMs). Such models use the dynamics equations and various physical parameterizations to compute the global circulation of the atmosphere in a self-consistent fashion. The models provide the three dimensional velocity field, pressures, temperatures, and various parameters related to cloudiness and precipitation on a uniform grid in longitude and latitude at several levels in the vertical at intervals of 10–30 minutes. This information can be saved and used as the input for a three-dimensional tracer continuity equation. In such a model there is no feedback from the chemical tracer evolution onto the general circulation. Rather, the transport is determined "off-line" using the output of the general circulation model. Such an approach allows for much longer integrations than are feasible if the calculation is done interactively as part of the climate simulation by the GCM. Of course for some problems, such as studies of the postulated "nuclear winter" resulting from global transport of smoke following a nuclear war, the effects of the tracer may greatly change the climate and an interactive approach is required (Malone et al. [48]). However, for gas phase chemical tracers the off line approach is quite satisfactory; only this approach is reviewed here.

Among the most extensive studies of global tropospheric tracer transport with such a model are those reported by Levy et al. [49] and Prather et al. [50]. The latter authors developed a chemical tracer model (CTM) that uses the dynamical fields output at 4 hourly intervals from an annual cycle simulation by the Goddard Institute for Space Studies (GISS) general circulation model.

In the first studies with the GISS CTM the fluorocarbons $CFCl_3$ and $CF_2Cl_2$ are used as the tracers. These are both synthetic molecules whose sources are almost entirely limited to the industrial areas of the Northern Hemisphere. They are very inert in the troposphere, and their distribution at the surface has been monitored by a number of stations from the Antarctic to the Arctic (see Fig. 28).

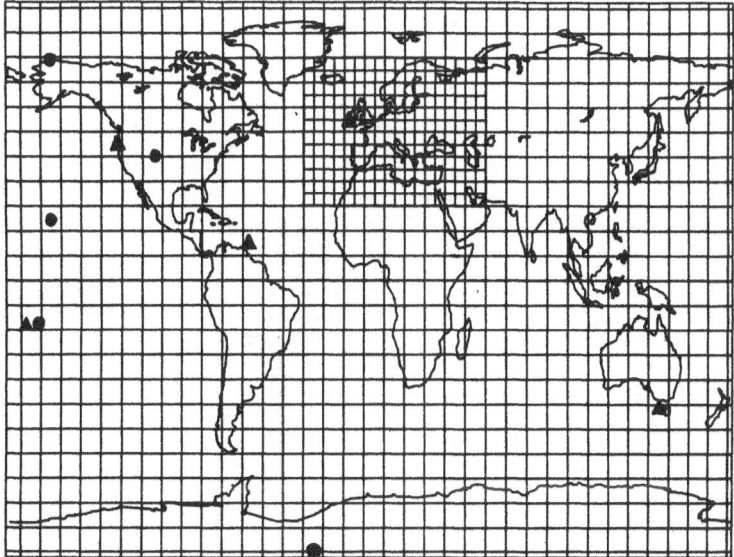

**Fig. 28.** Latitude – longitude grid network of the global chemical tracer model (CTM) showing locations of observing stations. The coarse resolution version of the model is shown except in the European sector where the fine mesh version is depicted. From [50]

An early finding of this research was that the interhemispheric transport as computed from the winds of the parent general circulation model was much too slow so that the difference between Northern and Southern Hemisphere concentrations was greater than observed. Prather et al., argued that the missing transport is provided in nature by mesoscale horizontal circulations associated with tropical convective storm systems that are not resolved by the general circulation model. They chose to parameterize this process by adding a horizontal diffusion term proportional to the local frequency of convective events in the model. The resulting surface and 50 kPa distributions of $CFCl_3$ are shown in Fig. 29. In both panels the observed differences between the Northern and Southern Hemisphere concentrations are well simulated. The surface map shows, not surprisingly, continental

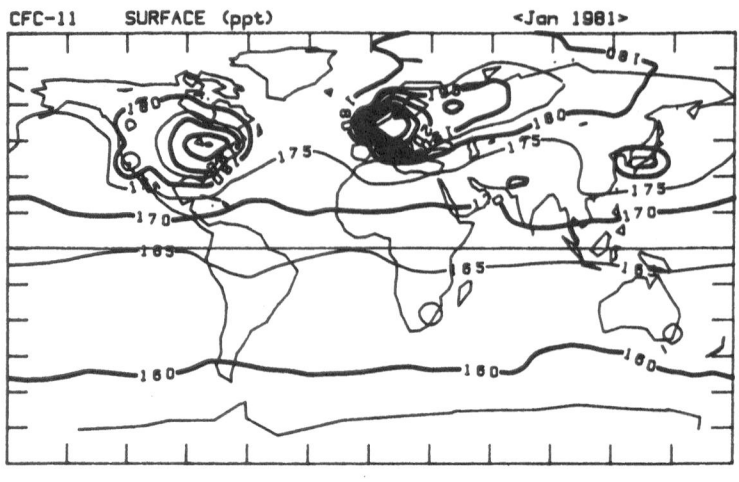

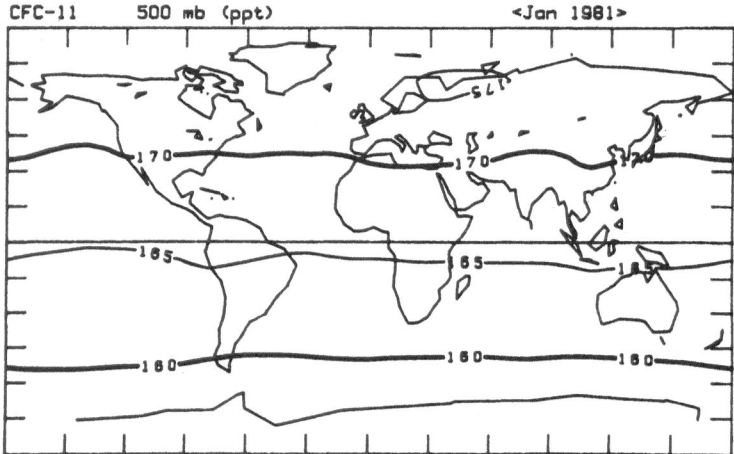

**Fig. 29.** Contours of CFCl$_3$ mixing ratios at the surface and 500 mb averaged for the month of January as determined by the CTM. From [50]

maxima over Europe and Eastern North America, corresponding to the locations of industrial emissions. But, these longitudinal dependencies are nearly absent in the mid-troposphere, showing that the east-west winds are very effective in homogenizing long-lived tracer distributions in the zonal direction in the middle troposphere. The distribution in the meridional plane (Fig. 30) indicates that in the troposphere significant vertical gradients occur only in the Northern Hemisphere where surface sources are strong. By the time this tracer has reached the Southern Hemisphere vertical mixing has eliminated these gradients so that the mixing ratio gradient in the Southern Hemisphere is nearly entirely in the meridional direction.

Because the CTM provides estimates of the tracer content in each grid box with a 4 hour time resolution it can also be used as an aid in interpretation of time series of

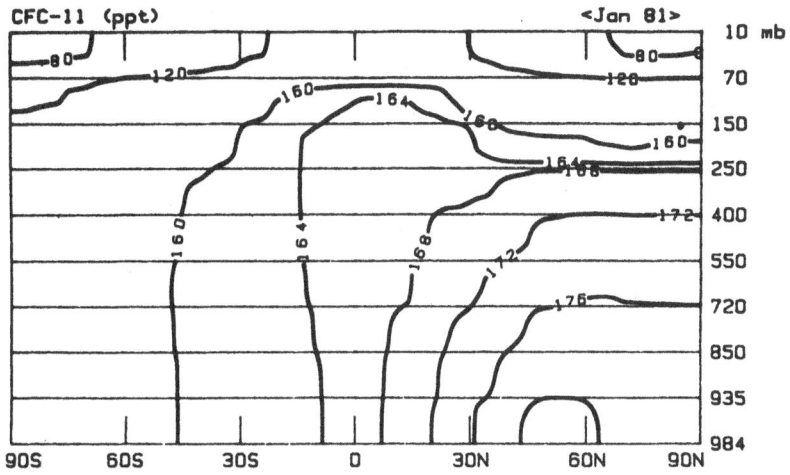

**Fig. 30.** Meridional cross-section of CFCl$_3$ mixing ratio surfaces in the CTM for January conditions. From [50]

observations. Fig. 31 shows the time series for 3 years of observations of CFCl$_3$ at Adrigole, Ireland. Superposed on the slowly increasing background value are sporadic occasions on which large positive deviations are recorded. Approximately 15 such pullution episodes are recorded each year associated with weather disturbances that transport continental European air to the Adrigole station. A time series for the same location produced by the CTM (not shown) exhibits a very similar pattern of tracer variability. Thus, the CTM can be used to analyze not only the mean climatology of globally distributed tracers, but the temporal variability for various locations around the world as well. Such information can be useful, for example, as a guide to the interpretation of tracer observations at a given location.

Extensive three-dimensional modeling of stratospheric transport has been carried out at the Geophysical Fluid Dynamics Laboratory (GFDL) in Princeton, New Jersey, USA. The GFDL tracer model is run off line using six hour averaged wind and pressure fields derived from annual cycle integrations of the GFDL general circulation model [51]. This model was used by Mahlman et al., [39] to elucidate the transport of long-lived trace gases with tropospheric sources, using N$_2$O as a characteristic example. The general pattern of the resulting tracer mixing ratio isolines in the meridional plane is in agreement with observations (Fig. 17); there is a tropical bulge and downward slope of the isolines toward higher latitudes.

Mahlman et al., examined the dependence of the mean tracer distribution on the chemical timescale by running simulations with 3 different specified chemical relaxation rates. The major differences between the slow and fast chemistry cases are in the vertical and meridional gradients of tracer mixing ratio in the stratosphere. These are much steeper for the short chemical timescale case, which is consistent with the more rapid loss of N$_2$O in that case. The slopes of constant mixing ratio surfaces in the meridional plane are, however, nearly the same in these two cases. Mahlman et al., suggest that this similarity should hold for all tracers in

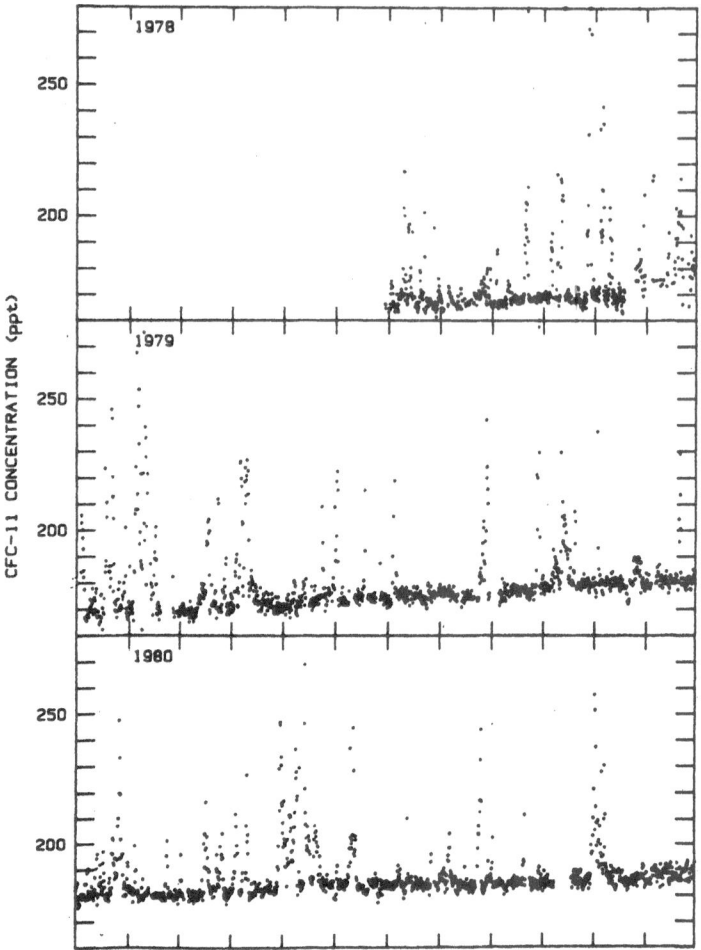

**Fig. 31.** Time series of individual observations of CFCl₃ from Adrigole, Ireland. From [50]

which the chemical timescale is long compared to the timescale for transport. For such tracers the mean slope results from a balance between the slope steepening tendency of the mean meridional circulation with its upwelling in the tropics and downwelling at high latitudes, and the slope flattening effect of meridional eddy transport and photochemical destruction. When eddy transport dominates tracer slopes will be similar for all tracers.

## Conclusion

In this review we have emphasized global transport in the meridional (height-latitude) plane, and mostly have restricted the discussion to the zonally averaged tracer budgets. Tracer gradients in height and latitude are generally much stronger than the longitudinal gradients. Of course, departures from zonal symmetry are crucial to the eddy transports that partly determine the zonal mean tracer distribution. Nevertheless, at the present stage of understanding it is reasonable to focus primarily on the zonally averaged view of transport.

Zonally averaged transport processes in the troposphere and stratosphere are summarized schematically in Fig. 32. Transport in the troposphere involves meridional advection by the mean Hadley circulation (much stronger in the winter hemisphere than in the summer hemisphere), rapid local transport by convective systems, and quasi-horizontal transport by transient synoptic and planetary scale eddies. Vertical transports and horizontal transports within a single hemisphere are quite rapid, so that in the absence of strong localized sources, long-lived tracers become well mixed within a few weeks. Interhemispheric transport is, however, much slower. A timescale of about a year is required for mixing between the hemispheres. Many long-lived tracers with unequal sources in the two hemispheres exhibit significant interhemispheric differences in concentrations.

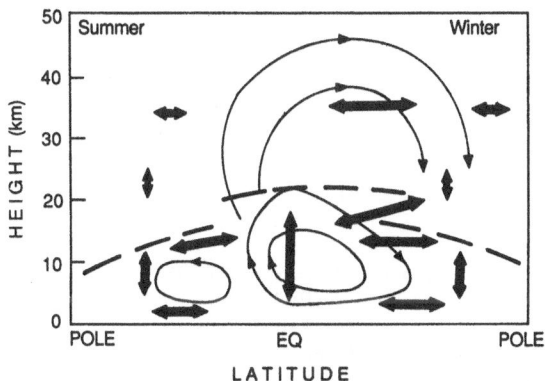

**Fig. 32.** Schematic illustration of transport processes in the troposphere and stratosphere. Heavy vertical arrows show vertical mixing by convection and diffusion. Heavy quasi-horizontal arrows show transport by large-scale eddies. Thin streamlines show the mean diabatic circulation. Dashed lines indicate the tropopause. Adapted from [3]

In the stratosphere vertical mixing by turbulent motions is very small and, except in the lowest layers, mixing by transient synoptic eddies is also small. Stratospheric transport is nearly entirely controlled by planetary scale eddies. Such eddies directly transport tracers meridionally in a quasi-isentropic fashion, and indirectly cause mean vertical and meridional transport through the diabatic circulation that arises in response to the radiative heating and cooling caused by eddy heat transports that drive the temperature field away from its radiatively determined distribution. Current estimates [3] suggest that the time scale for turbulent mixing over one scale height in the stratosphere is a few years, while the time scale for upward advection by the diabatic circulation is a few weeks except near the tropopause where it becomes several months. The time for meridional mixing ranges from more than a season in the summer hemisphere, to order of a few days during periods of enhanced wave activity in the winter hemisphere. Overall the equator to pole transport time in the stratosphere is probably about 100 days, and so is comparable to the time scale for the seasonal cycle of the global circulation.

*Acknowledgement*: I wish to thank Dr. Michael Prather for helpful suggestions.

## References

1. Palmen E, Newton C W (1969) Atmospheric circulation systems. Academic, New York
2. Holton J R (1979) An introduction to dynamic meteorology. Academic, New York
3. WMO (1985) Ozone Assessment Report, 1985. World Meteorol. Organ., Geneva
4. Labitzke K (1982) J. Meteorol. Soc. Jpn. 60: 124
5. Farman J C, Gardiner B G, Shanklin J D (1985) Nature, 315: 207
6. Jukes M N, McIntyre M E (1987) Nature, 328: 1
7. Solomon S (1988) Reviews of Geophysics, 26: 131
8. Andrews D G, Holton J R, Leovy C B (1987) Middle atmosphere dynamics. Academic, New York
9. McIntyre M E (1980) Philos. Trans. R. Soc. London, Ser. A296: 129
10. Welander P (1955) Tellus 7: 141
11. Mahlman J D, Andrews D G, Hartmann D L, Matsuno T, Murgatroyd R J (1984) Transport of trace constituents in the stratosphere. In: Holton J R, and Matsuno T (eds). Dynamics of the middle atmosphere. Terrapub, Tokyo
12. Andrews D G, McIntyre M E (1978) J. Fluid Mech., 89: 609
13. Danielsen E F (1968) J. Atmos. Sci., 25: 502
14. Russell G L, Lerner J A (1981) J. Applied Meteor., 20: 1483
15. Gidel L T (1983) J. Geophys. Res., 88: 6587
16. Prinn R G, Simmonds P G, Rasmussen R A, Rosen R D, Alyea F N, Cardelino C A, Crawford A J, Cunnold D M, Fraser P J, Lovelock J E (1983) J. Geophys. Res. 88: 8353
17. Craig H, Chou C C (1982) Geophys., Res. Lett., 9: 1221
18. Logan J A, Prather M J, Wofsy S C, McElroy M B (1981) J. Geophys. Res. 86: 7210
19. Logan J A (1985) J. Geophys. Res., 90: 10463
20. Telegadas K, List R J (1969) J. Geophys. Res., 74: 1339
21. Brewer A W (1949) Quart. J. Roy. Meteorol. Soc., 75: 351
22. Dobson G M B (1956) Proc. Roy. Soc. London, A129: 411
23. Houze R A (1982) J. Meteorol. Soc. Jpn. 60: 396
24. Kley D, Schmeltekopf A L, Kelly K, Winkler R H, Thompson T L, McFarland M (1982) Geophys. Res. Lett. 9: 617
25. Newell R E, Gould-Stewart S (1981) J. Atmos. Sci. 38: 2789
26. Atticks M G, Robinson G D (1983) Quart. J. Roy. Meteorol. Soc., 109: 295
27. Danielsen E F (1982) Geophys. Res. Lett., 9: 605
28. Johnson R H, Kriete D C (1982) Mon. Wea. Rev., 110: 1898
29. Shapiro M A (1980) J. Atmos. Sci., 37: 994
30. Dopplick T G (1979) J. Atmos. Sci. 36: 1812
31. Holton J R (1986) J. Geophys. Res., 91: 2681
32. Schmidt U, Kulessa G, Khedim A, Knapska D, Rudolph J (1984) ESA Symp. Eur. Rocket Balloon Programmes, 6th ESA SP-183: 141
33. McIntyre M E, Palmer T N (1983) Nature 305: 593
34. Jones R L (1984) Adv. Space Res. 4(4): 121
35. Bowman K P, Krueger A J (1985) J. Geophys. Res. 90: 7967
36. McPeters R D, Heath D F, Bhartia P K (1984) J. Geophys. Res. 89: 5199

37. Siegenthaler U, Oeschger H (1987) Tellus 39B: 140
38. Harte J (1985) Consider a Spherical Cow: A Course in Environmental Problem Solving. Kaufmann, Los Altos, CA
39. Mahlman J D, Levy H I I, Moxim W J (1986) J. Geophys. Res., 91: 2687
40. Rodhe H, and Isaksen I S A (1980) J. Geophys. Res., 85: 7401
41. Hyson P, Fraser P J, Pearman G I (1980) J. Geophys. Res. 85: 4443
42. Pearman G I, Hyson P (1986) J. Atmos. Chem., 4: 81
43. Isaksen I S A, Hov O (1987) Tellus 39B: 271
44. Reed R J, German K E (1965) Mon. Wea. Rev., 93: 313
45. Ko M K W, Tung K K, Weinstein D K, Sze N D (1985) J. Geophys. Res., 90: 2313
46. Garcia R R, Solomon S (1983) J. Geophys. Res., 88: 1379
47. Solomon S, Garcia R R (1984) J. Geophys. Res., 89: 11633
48. Malone R C, Auer L H, Glatzmaier G A, Wood M C, Toon O B (1986) J. Geophys. Res., 91: 1039
49. Levy H I I, Mahlman J D, Moxim W J (1982) J. Geophys. Res., 87: 3061
50. Prather M J, McElroy M B, Wofsy S C, Russell G, Rind D (1987) J. Geophys. Res., 92: 6579
51. Mahlman J D, Levy H I I, Moxim W J (1980) J. Atmos. Sci., 37: 655

# The Atmosphere: Physical Properties and Climate Change

*Reiner Eiden*

Universität Bayreuth, Lehrstuhl für Hydrologie, Abteilung für Meteorologie,
D-8580 Bayreuth, Federal Republic of Germany

## Summary

This contribution discusses the known facts and possible causal mechanisms of global climatic change. After starting with some remarks on the problem of defining and describing climate the subject is presented in three main parts. The first deals with the acquisition of climate data and with the data themselves. Different data sources are specified. Emphasis is put on isotope ratios of datable strata of deposits and sediments—preferably layered ice cores—which increasingly serve as a detailed source of climate information. Time-series particularly of mean annual surface temperature, at present representing the most reliable climate element, are itemized. The second part is devoted to the causes of climate change. The known natural as well as anthropogenic causes are discussed. The third part introduces some basic features of climate modelling. Included are new modelling experiments and some results of special studies on possible changes and trends of present atmospheric conditions.

## Introduction

In the early days of atmospheric research, climatology—the science of the longterm behavior of the atmosphere—was a descriptive almost historical kind of science oriented to the past. The emphasis on discription demonstrates our initial inability to observe and to record atmospheric conditions and to discern the atmospheric causes and trends behind the reports of the disasters, including famine, flood, drought, and disease. With accumulating knowledge, progress in observational, experimental, and theoretical methods, with an increasing variety of data sources and time series of climatic data the longlasting conditions, fluctuations and changes of the atmosphere became evident. In reconstructing the climate of the past, the course of the future climate also became a topic of interest, primarily for general speculation according to the state of knowledge at the time. With the rapid advances in our understanding of atmospheric processes over the past 15 to 20 years, the physical, chemical, and computational basis broadened to the forecast of climate. In recent years it has become increasingly clear that the climate is subject not only to natural influences but also to anthropogenic activities. This has led to a high motivation to study, and to model the complex mechanisms which govern climate and the environment as a whole.

These studies concentrate more or less on climate variations caused by the actual changes in the physical and chemical composition of the atmosphere. Current efforts in particular are trying to develop models that will produce reliable estimates of the response of known or possible human activities. Paramount among these is the anticipated change due to the carbon dioxide content which has been increasing globaly since the turn of the century. On theoretical grounds, an increase of $CO_2$ or other infrared absorbing gases which also intensify the atmospheric greenhouse effect should lead to an appreciable warming of the surface air. A possible reason for the failure of the expected general warming arises from the possibility of negative, damping feedback mechanisms due to changes in the aerosol particle content and cloudiness. The physical properties of atmospheric aerosol particles play a critical role in this context. Numerical models have been developed which enable the description and the inclusion of the components of the global climate system to be made the interaction of the atmosphere with the oceans, ice and land masses of the earth and the consideration of the energy and material (especially water) exchange between these different components. Research concerned with astronomical variations, past and local climate events is at present less developed than the development of sophisticated numerial models. The discussion of the present state of art in this article is far from being comprehensive. Knowledge of climatic processes is rapidly evolving and simultaneously expanding; and adjacent branches of science are becoming more and more involved. Biological and socio-economic effects, which may be caused by the changing climate for instance, are increasingly treated by the scientific community in addition to the primary physical and chemical effects of changing atmospheric conditions. Such secondary effects, however, will not be discussed here.

## Climatic System, Definition, and Elements

The physical condition of the atmosphere or "the weather" is the result of heating and cooling of the planet Earth by solar and terrestrial radiation and the consecutive reactions—circulation of the atmosphere and oceans and the transfer of latent and sensible heat. Climate means the synthesis or long-term manifestation of weather (Durst [24]). It is, in general, represented by the statistical collective of some characteristic atmospheric variables documented by time-series of compatible specified length. The variables are called the climatic elements; the length of climatic series is by international convention 30 years: the reference or normal period [109]. In 1956, the World Meteorological Organization (WMO) recommended the use of the most recent available period of 30 years, starting on 1 January of a year ending with the digit 1 (e.g. 1951–1980). A climatic element may be any property which specifies the physical state of the atmosphere. The classical elements are air temperature, amount of precipitation, pressure, wind and duration of sunshine.

This more or less intuitive, simple definition of climate, considering only the long-term mean conditions of the atmosphere, obstructs an approach to the understanding of the climatic variations. Modern climatology, for instance, trying to forecast the long-term conditions of the atmosphere has, in addition, to take into account the atmospheric boundary conditions and its variations [69]. The atmosphere is not an isolated physical or physico-chemical system. It is linked to other "spheres" and it interacts with them [13]. Such interrelations are very complex and are not yet completely understood. They are the subject of present research activities.

The climatic system with its subsystems and elements is shown in Fig. 1. Two different principal processes can be distinguished in climate formation and climatic change:

internal processes—interactions between the components of the system: atmosphere, cryosphere, biosphere, hydrosphere, lithosphere

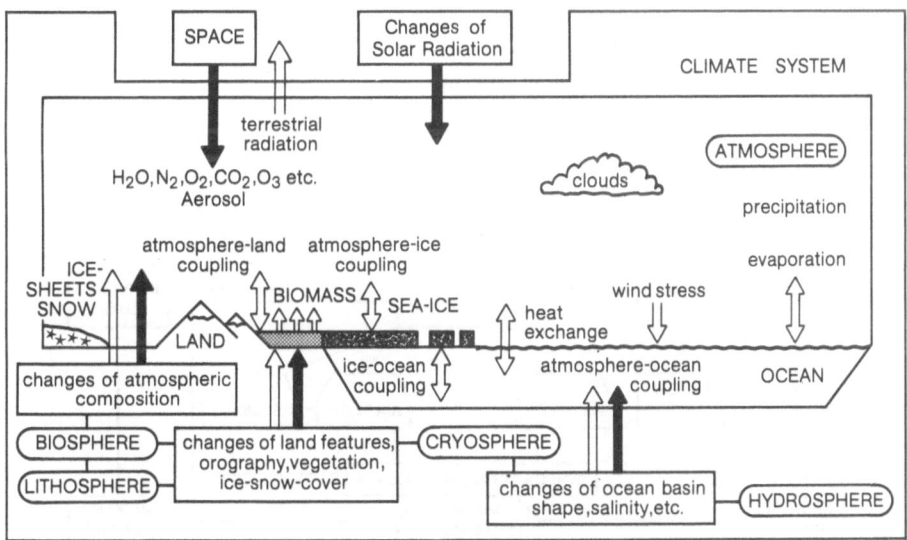

**Fig. 1.** Schematic illustration of the climate system. External processes are indicated by full arrows, open arrows indicate internal processes. (From US GARP Committee 1975 [128], modified)

external processes—unidirectional actions influencing the system from outside or originating in the system itself.

The internal climate subsystems, above all the atmosphere, can be characterized by relatively rapid fluctuations, i.e. the weather. Whereas the external climate subsystem provides relatively slow changing external influences on the internal system. "Climate, including its changes, may then be defined in terms of averages over a hypothetical ensemble of internal states that is nearly in equilibrium with the slowly changing external influences: climate changes as the external conditions change" (Schneider and Dickinson [108]).

Variables of external processes are solar radiation, particularly the long-term changes in the solar constant, insolation and terrestrial orbit; meteorite dust; salinity of oceans; volcanic activity; and anthropogenic activities. A change of salinity for instance causes a change in the concentration and the physical properties of condensation nuclei which results in a change of cloudiness. Meteoric dust, anthropogenic, and volcanic aerosol particles immediately influence the radiation transfer within the system.

There are a large number of internal or feedback processes which occur within the system. Some may act to amplify the variations (positive feedback) while others act to dampen them (negative feedback). The ice–albedo interaction represents a positive feedback; the high solar reflectance of an ice or snow cover leads to further cooling of the surface. The water vapor-radiation feedback is also a positive process; an increase of water vapor intensifies the atmospheric greenhouse effect. This will lead to an enhanced surface temperature resulting in an enhanced evaporation. An example of a negative feedback is the cloud-radiation interaction. Increased cloudiness will reduce the amount of solar radiation reaching the earth surface,

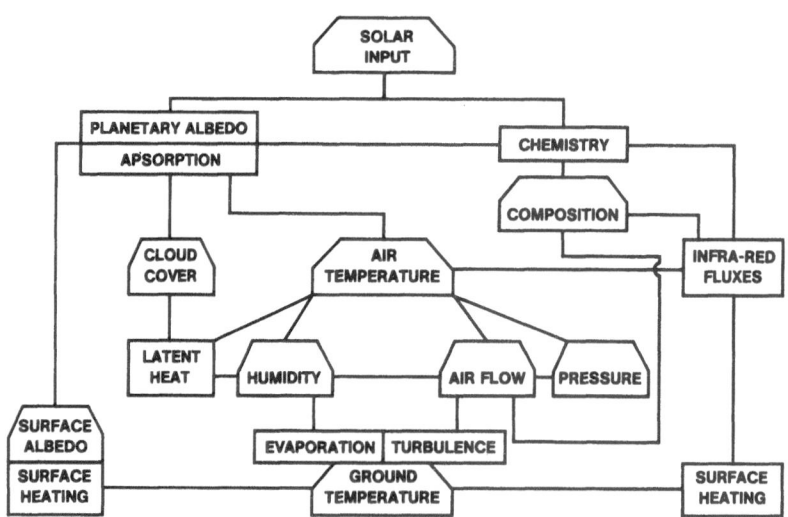

Fig. 2. A schematic box representation of the interdependence of some climate variables ( ⬭ ) and processes ( ▭ ) responsible for climate and climatic change. (From Davies 1985 [20], modified)

which will tend to reduce the evaporation of water vapor and the forming of clouds. A schematic illustration of some at these interrelations is shown in Fig. 2.

To develop a satisfactory understanding of the physics of climate, and the past and future climate variations we need information on more than just the classic climatic elements. Some further elements for which we need continuous time series are: solar radiation, evaporation from land and sea surfaces, and snow and ice cover. Presently we do not even reliably know all of the few classic elements of the recent historical period.

The time scale of the various processes of the subsystem atmosphere ranges from a fraction of a second (turbulence) to millennial events (ice ages). The atmospheric process "climate" is situated at the long end of this time scale starting with climatic fluctuations lasting a few years [144].

### Branches of Climatology

The structure of the climatic system and the nature of the different spheres clearly show that climatology is an interdisciplinary science. Consequently, different research fields have developed depending on special questions and interests. The main branches, especially those in the process of formation or newly expanding are listed below.

Bioclimatology deals with the relations of climate and life, especially the effects on the growth conditions of plants and animals as well as the health and activity of human beings. Subbranches are agricultural climatology, ecology, human and medical bioclimatology [31, 126, 146]. The forecasting of social and economic conditions based on the future climate is part of this branch [64].

Technical climatology is concerned with the interrelation between climate and engineering projects. Examples are city planning, highway, and building construction [77, 79]. The activities of the weather service may also be seen within this frame. Subbranches are city climatology, aviation climatology, synoptic climatology [66, 103].

Physical climatology is the major branch of climatology. It deals with the description of climate on the basis of physical and physico-chemical laws and processes, rather than simply with an empirical presentation. Subbranches are dynamic climatology, climatology of the free atmosphere, radiation climatology [10, 117, 145].

Paleoclimatology studies the "prehistoric" climate or the climatic changes in geological times. It is a branch of geology and it is only mentioned here for completeness.

Depending on the scale, we can differentiate between microclimate, for instance in the environment of a blade of grass, mesoclimate, for instance in the urban environment, and the macroclimate, for instance the climate of a hemisphere or the global climate. Because of their general importance and since the change of the macroclimate will also influence the minor scaled systems, the following sections deal primarily with macroscale effects.

## Sources of Climate Data

The foremost source of evidence for climatic changes is the instrumental record. This source, however, only provides data for modern climate history. The study of isotope ratios of natural substances gives, with increasing success, the most comprehensive indirect information on past climates. The documentary evidence of proxy data ranges with changing reliability between these two sources; proxy data embrace any material that provides an indirect measure of climate—e.g. historical reports, vine harvests, tree rings, snow lines, movement of glaciers etc. A summary of the numerous sources of proxy data is given by Lamb (1982 [61]). The isotope data formerly included in the proxy data group are now treated as a separate class because of their advancing physico-chemical exactness and reliability.

### Instrumental and Proxy Data

Meteorological instruments, suitable for continuous measurements, have been available for the last 300 years. The barometer and thermometer were both invented in the 1640's. A rain gauge was used in Lancashire, England, in 1676 and soon after that in various other places. The first known longtime wind observations originate from London starting in 1670. Western Europe generally has the longest, continuous records (Lamb and Johnson [63], Tooley and Sheail [124], Manley [76]). A detailed review is given by Von Rudloff [104]. The famous so called "Hundertjährige Kalender" is based on weather observations by Abbot M. Knauer in Central Germany from 1652–1658. Thereafter these observations have been repeatedly joined together covering a time period of 100 years [28].

Instrumental records from an organized international global network of climatological stations has been available only for the last 100 years. A first attempt to arrange such a network with a standardized instrumentation was made by the "Societas Meteorologica Palatina" (Mannheim, Germany) in 1781. This first synoptic network comprised 39 stations and was successfully operated up to 1795. Upper-air data span only about 40 years. The world wide and continuous measurement of atmospheric parameters such as solar radiation, $CO_2$, $O_3$, dust content, snow cover is largely of more recent origin [96]. The problems involved in using the readings of early instruments are discussed by Mitchell et al. [83].

Historical reports, an undoubtedly interesting source of information on weather and climate, will only be briefly mentioned here. A comprehensive documentation is given by Le Roy Ladurie [70] and by Lamb [60]. Records of bad harvests, devastating famines, quality of wine, wheat prices, weather catastrophes, glacier extent, sea-ice conditions, taxation revenues etc. give evidence of sometimes tragic climatic processes. The decline of the Greenland Old Norse colony after 1400 may illustrate this.

Biological, geological and glacial processes sensitive to the conditions of the surrounding atmosphere preserve, once terminated, the environmental conditions. They yield secondary data of the past. The interpretattion of such "proxy" and in general complex data in terms of climatic elements is difficult and require a critical assessment [32, 140]. Yet they provide valuable information and time series of climatic elements for the time before instrumental record and overlapping it. Three sources of this kind will be itemized here without discussing their methodology.

The dendroclimatology is concerned with the analysis of the wood-growth layers in plants or tree rings and the extraction of climatic information. It is based on the knowledge of plant growth depending on climatic conditions preceding or during the growing season. This dating method yields the most reliable results for instance of tree ring-samples taken at semi-arid or warm sites, where ring widths vary with the intensity and duration of drought or of tree growing in high-altitude or high-latitude sites where the temperature of the growing season is the dominant parameter [33, 34]. The method can be extended to large-scale climatic controls and it yields the possibility of reconstructing seasonal conditions for individual years and sites. So the potential use of frost rings in trees to reconstruct climatically effective volcanic events has been studied and proved by LaMarche and Hirschboeck [59]. Most recently Lough and Fritts [71] used tree-ring data to reconstruct the annual average temperature of North America and the cooling by volcanic events within the period 1602 to 1900 A.D. A further improvement of the method is very promising since individual ring-width chronologies extend more than 8000 years BP [60, 96].

Glaciology complements the results of dendroclimatology in that it relates primarily to high-latitude and high-altitude sites [93]. In general, glaciology implies the study of snow and ice on the earth's surface, with a specific concentration on the field of active glaciers. Glacial motions are strongly determined by the vertical mass-elevation profile of the glacier. An advance may signify colder summers but also, in high latitudes, increased snowfall which is usually associated with warmer winters. To extract more unique climatic information, models have to be used which compute the possible combinations of climatic conditions that could produce the observed changes of the glacier or snow line. An example for such a procedure may be the energy-mass balance model for a glacier, developed by Williams [141, 142]. A selection of appropriate references is given by Untersteiner [127].

Palynology, the study of the pollens of seed-bearing plants, contribute not only to our perception of the past climatic record but also to our perception of plant migration caused by a climatic change [3, 6]. Pollens and also other macro- and microfossils are preserved in waterlogged environments, such as peat bogs and lake sediments, where lack of oxygen prevents the usual decay. The analysis of the pollen spectrum and the stratigraphic pollen profile of undisturbed peat or lake deposits, yield long time series with comparatively long stretches of time during which the composition of the vegetation remained substantially unchanged and rapid changes whenever they occur; climatic fluctuations are smoothed out whereas true climatic changes are well documented. An analysis of the pollen spectrum of the Sudeten mountains for instance clearly show the decline of the upper tree line by 100 to 200 m with the onset of the little ice age [30].

### Isotope Ratios of Natural Substances

A powerful tool for acquiring climatic data of the past is represented by the analysis of isotope ratios. The method was worked out first by Urey (1947), further developed and widely used by Dansgaard and associates [18, 19] and first applied to oxygen. The isotopes $^{16}O$ and $^{18}O$ are suitable for such an analysis. The isotope $^{17}O$ is a minute component and can be neglected. They are present in water in all its phases forming the most important isotope components $H_2^{16}O$ and $H_2^{18}O$. Their average concentrations in the world's oceans are 99.759 and 0.204% respectively. There is a fractionation depending on temperature when water passes from one phase to another or when oxygen is exchanged between different substances. The vapour pressure of the heavy isotope $H_2^{18}O$ is less than that of $^1H_2^{16}O$ and the difference decreases with increasing temperature. So when water evaporates from oceans the vapor will become more depleted in $^{18}O$. The first condensate will still tend to have almost the same oxygen ratio as the ocean water. As the water vapor is transported to colder regions of the atmosphere, i.e. higher altitudes and higher latitudes, it will become progressively more depleted in $^{18}O$, since the difference of the vapor pressure between $H_2^{18}O$ and $H_2^{16}O$ increases with decreasing temperature. Thus by increased cooling the air mass carrying the water vapor gives off precipitation whose $^{18}O$ content decreases as the temperature of the air mass decreases. The precipitation is depleted in $^{18}O$ and isotopically lighter. Consequently the isotope ratio of a rain drop or an ice crystal is primarily a function of the mean temperature of the surrounding air during his history and formation and prior to its deposition on the earth surface. Similar processes occur during the formation of the shells of biological organisms in the ocean. The $^{18}O/^{16}O$ ratio of the carbonate $(CaCO_3)$ of the shell of the planktonic species for instance, which is built up near the ocean surface by carbon dioxide $(CO_2)$ and the calcium hydrogen carbonate solute $(Ca(HCO_3)_2)$, indicates the temperature of the surrounding water layer near the ocean surface. The shell of the benthonic species indicates the temperature near the sea bed. A stratigraphic analysis of an ocean sediment core composed of dead organisms [94] or of ice cores from polar ice sheets provides a long-term temperature record.

The continuous profiles of isotopic composition along the cores are given by the "del" or $\delta$ function (per mil)

$$\delta^{18}O = \frac{{}^{18}O/^{16}O_{\text{(sample)}} - {}^{18}O/^{16}O_{\text{(SMOW)}}}{{}^{18}O/^{16}O_{\text{(SMOW)}}} 10^3 \ 0/00$$

This function $\delta^{18}O$ measures the $^{18}O/^{16}O$ isotope ratio of a sample relative to the corresponding isotope ratio of the "standard mean ocean water" (SMOW) [15].

The $\delta^{18}O$ value of the precipitation is negative. It can be related to latitude or furthermore to the local mean temperature at the place of precipitation. The measured $\delta$ values of plant material are positive. The mechanisms which determine the $\delta$ function of the plant material are very complex. They include a temperature-dependent isotope fractionation associated with the metabolic reactions occuring during photosynthesis and with the evapotranspiration of leaf water [40]. Figure 3 shows the relation between the annual mean of the $^{18}O$ content of the cellulose of different plants and the mean local surface air temperature or geographical latitude. The temporal variation of $\delta$ measured at a definite place reflects the seasonal and, depending on the age of the sample, also the climatic temperature conditions of this

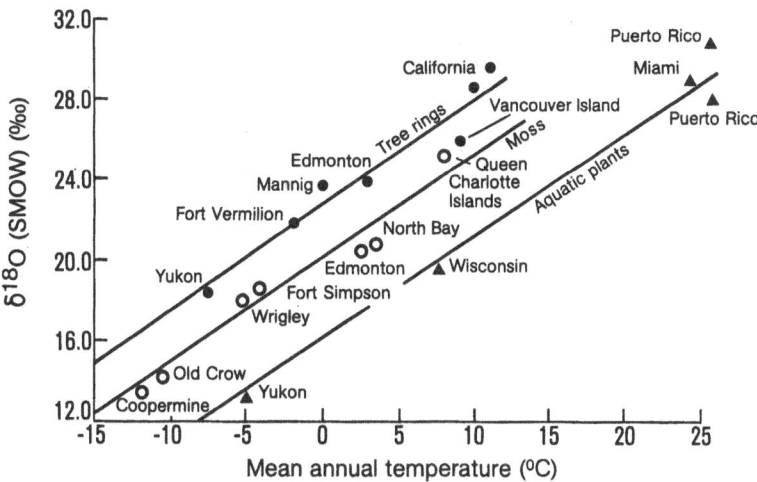

**Fig. 3.** Dependence of cellulose $\delta^{18}O$ values on mean annual temperatures at various sites: ● tree rings, ○ sphagnum moss (peat producing plants), ▲ aquatic plants. (Redrawn from Gray 1981 [40])

place. The measurement of the variations in the isotope ratio is done by isotope mass spectrometry.

Untill now the study of the oxygen isotope ratio has supplied most results. This may be expected since a large reservoir of oxygen exists in the hydrosphere; furthermore, oxygen is very reactive and forms compounds with most other elements. Significant climatic information, however, can also been obtained from other stable isotope ratios. The greater the mass difference between two isotopes in relation to their absolute mass, the greater will be the isotope effect. So studies concentrate increasingly on low atomic mass elements: carbon $^{13}C/^{12}C$, nitrogen $^{15}N/^{14}N$, and particularly deuterium–hydrogen D/H ($\delta$D) but also sulphur $^{34}S/^{32}S$ [40]. A paper recently published by Aristarian et al. (1986 [2]) represents a first attempt to reconstruct the antarctic temperature variation of the past beyond available instrument records. It is based on a deuterium analysis of an ice core sample taken from James Ross Island (64° 13′ S, 57° 40.5′ W, alt. 1640 m a.s.l.). The study covers the past 130 years (1850–1980) and presents the temperature trend for this period—at least of the antarctic region. The detailed analysis of the deuterium $\delta$D (SMOW) profile of the ice core and the temperature data of the neighbouring stations (e.g. Orcadas island, 1904–1980) shows a strong correlation of the extreme values with a shift of one or two years. The data represented in Fig. 4 differently low pass filtered and smoothed do not clearly reflect this result.

The climatic data extracted from the Vostok ice core published 1987–88 already cover a period of 160,000 years [4, 54, 68]. The 2.083 m ice core was recovered by the Soviet Antarctic Expedition at the research station Vostok (East Antarctica, 72° 28′ S, 106° 48′E, alt. 3490 m a.s.l). It fully covers the last glacial-interglacial cycles. Besides data on temperature, data on atmospheric carbon dioxide and aerosol particles have been obtained [Fig. 5]. The data are interpreted on the basis of the deuterium profile $\delta$D (SMOW) though the ice core has been analyzed for the

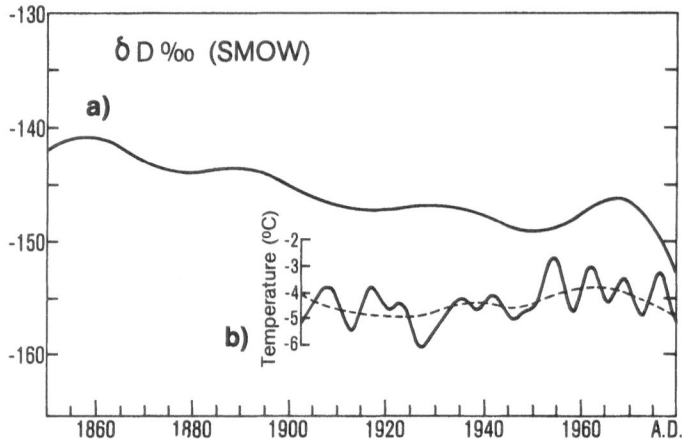

**Fig. 4.** Antarctic temperature trend (1850–1980)

**a)** mean annual deuterium $\delta D$ profile versus time smoothed using a 13 year low-pass filter; ice core samples drilled at Dalinger Dome James, Ross Island (West Antarctica, 64° 13′ S, 57° 40.5′ W)

**b)** mean annual surface temperature versus time, data 3 year low-pass filtered; instrument record from Orcadas Island (60° 45′ S, 44° 43′ W) starting 1903. (Redrawn from Aristarain et al. 1986 [2])

oxygen content too. Deuterium turns out to be a slightly better indicator of temperature change [54].

Air bubbles occluded in the pores of the ice sheets and isolated from the surrounding atmosphere provide important, reliable information on atmospheric $CO_2$ particularly on the pre-industrial $CO_2$ level and the recent increase caused by anthropogenic activities.

Previous to the Vostok study, Neftel et al. [84,85] reported the extraction of trapped air and the determination of $CO_2$ by IR-laser spectrometry. The measurements reflect the atmospheric $CO_2$ content during the past 40,000 years. In the frame of this investigation, ice core samples recovered in Greenland and the Antarctic have been analysed. The study indicates, probably due to atmospheric mixing, a fairly good synchronization of the $CO_2$ concentrations of the Northern and Southern Hemispheres. This is at least qualitatively confirmed by the measurements. (A more generalized representation of the variation of $CO_2$ given by Neftel et al. [85] is shown in Fig. 17b.) The corresponding $CO_2$ data for the past 40,000 years, derived from the Vostok deep ice core (compare Fig. 5b) are inserted in this figure. These antarctic data are near the lower limit of the $CO_2$ range estimated by Neftel et al. This supports the findings of Neftel et al. concerning differences between antarctic and arctic $CO_2$ concentrations observed. They pointed out that the extension of the data range to higher values seems to be mainly caused by arctic data; these data are probably increased and adulterated by meltwater.

A further study of the Vostok deep ice core also yields a comprehensive concentration profile of all major ionic impurities, which can be attributed to atmospheric aerosol particles. In Fig. 5c only the profile of the magnesium ion, representing substances of maritime origin, is depicted. This ion has also a

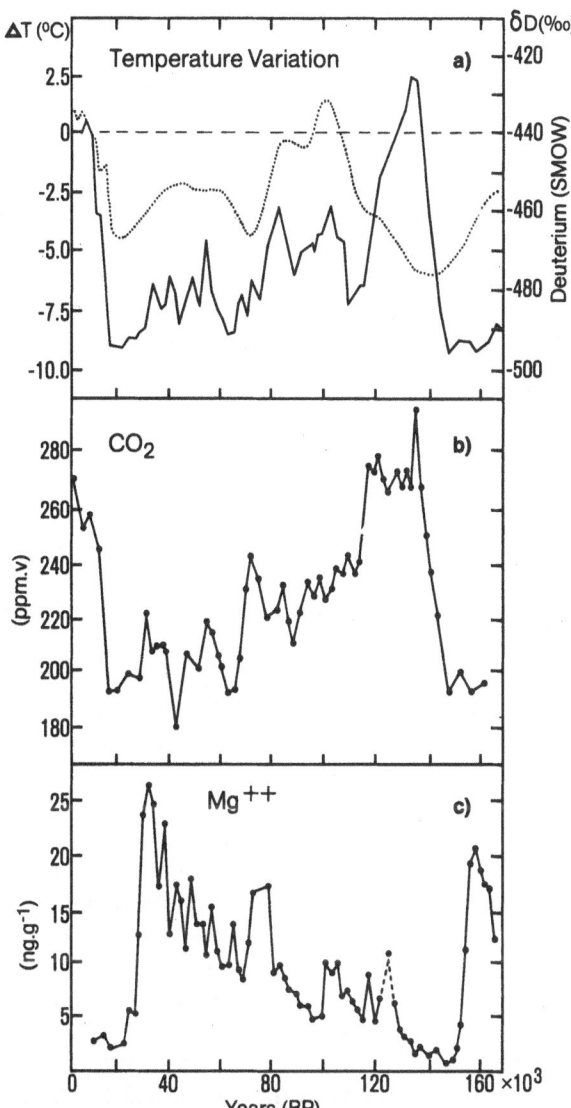

**Fig. 5.** Deuterium $\delta D$ profile of a 2083 m ice core drilled at the Soviet Antarctic Station Vostok (East Antarctica, 72° 28′ S and 106° 48′ E) covering a time period of 160,000 years (BP). Derived quantities:
**a)** Surface temperature variation; difference from modern mean annual surface temperature of the site ($-55.5\,°C$). The doted curve represents the estimated surface temperature variation of the northern hemisphere (compare Fig. 8)
**b)** Variation of mean atmospheric $CO_2$ surface concentration (best estimate)
**c)** Variation of Magnesium ions originating from marine aerosol and terrestrial dust (soluble impurity)
(Redrawn from Jouzel et al. 1987 [54], Barnola et al. 1987 [4] and Legrand et al. 1988 [68])

significant fraction of terrestrial origin and is linked to substances of continental sources. The climatic history of all the ions analyzed ($Na^+$, $NH_4^+$, $K^+$, $Ca^{2+}$, $H^+$, $Cl^-$, $NO_3^-$, $SO_4^{2-}$) show the same general behaviour: in contrast to $CO_2$ the concentrations of these ions increased during cold periods and decreased during warm periods. A detailed discussion of the results of this study as well as references on former studies is given by Legrand et al. (1988 [68]).

Datable layered ice cores also provide records of past precipitation. The accumulation rates can be deduced if it is possible to correct the layer thickness for flow effects within the ice sheet. A sophisticated method is proposed by Paterson and Waddington [90]. It is based on Berylium 10 ($^{10}Be$), which is produced by cosmic radiation within the atmosphere and which has a half-life time of $1.5 \times 10^6$ years. It becomes attached to aerosol particles and is at least partly deposited as condensation or ice nuclei incorporated in precipitation elements after a mean atmospheric residence time of about 1 year.

**Dating Problems**

To reveal climatic variations using isotope ratios, it is necessary to find natural systems which have recorded these variations in datable strata. According to Gray [40] such a system must satisfy a number of criteria: (a) the isotope variations found within the system must be reducible unambiguously to climatic factors; (b) after the strata have been deposited, the isotope record must be permanent and not exposed to further fractionation processes; (c) the record preserved by the strata should be continuous and datable so that a time scale can be attached to the record with a time resolution of the stratified system commensurate to the climate information required.

There are some natural systems known which, to a sufficient extent, satisfy these criteria: deep-ocean and lake sediments [94], speleothem deposits of $CaCO_3$, peat deposits, annual coral rings, annual growth rings of trees and polar ice sheets from regions with maximum temperatures well below freezing point, so that perturbating effects by percolation can be excluded. A sediment chronology of quaternary time scale distorted, for instance, by unknown changes in sedimentation rate can be corrected or tuned considering the known history of orbital forcing [78]. The procedure presupposes a linear climate response (e.g. of a continuous isotope stratigraphy) to variations in the earth's orbital geometry.

In order to date natural deposits of the last 10,000 to 50,000 years with the sufficient resolution, the use of the radiocarbon ($^{14}C$) technique plays an important role. Radiocarbon is produced by cosmic rays—mostly protons—penetrating the atmosphere from the galaxy. A part of these high energy particles form neutrons ($^1_0n$) in a primary process. When these neutrons collide with atoms of nitrogen, the atmosphere's main constituent, the dominant reaction is

$$^1_0n + {}^{14}_7N \rightarrow {}^{14}_6C + {}^1_1H.$$

The hydrogen atom so formed originates from the proton lost by the nitrogen nucleus capturing the neutron. The radiocarbon atoms produced become successively oxidized and form part of the atmospheric carbon dioxide.

The $^{14}CO_2$ incorporated in plants and animals by assimilation or consumption of assimilation products is in balance with the $^{14}C$ content of the atmosphere. This is true for the living organism. From the moment an organism dies the exchange with the atmospheric reservoir ceases. The decay of the radiation of $^{14}C$ in the organism is uncompensated, the radiocarbon watch starts to run down marking the time of death.

The radiocarbon technique carries intrinsic limitations, however. One reason is the rather short half-life value of $^{14}C$ on a geologic time scale of $5730 \pm 40$ years. By thermal diffusion isotopic-enrichment techniques the range of dating can be extended up to 75,000 years [41]. In addition to laboratory error, possible contamination of samples by modern organic matter, plant roots etc., changing atmospheric $^{14}C$ reservoir caused by changing cosmic radiation and sun activity have reduced the dating precision at least of samples of the last 6000 years. Besides a general error of $\pm 50$ years, Stuiver [123] has shown that specially during the last 450 years the radiocarbon dating is ambiguous; any value for a sample may indicate a number of possible calendar dates (Fig. 6). The proportion of radioactive carbon in the atmosphere's carbon dioxide is not as constant as it was assumed just after the technique was worked out by Libby in 1946.

On the other hand absolutely dated tree ring material from almost all continental areas is available. The bristlecone pine chronology from the White Mountains in California, for instance, has a well documented length of 7000 years (Ferguson [29]). This material can be used to get information about external climate variables,

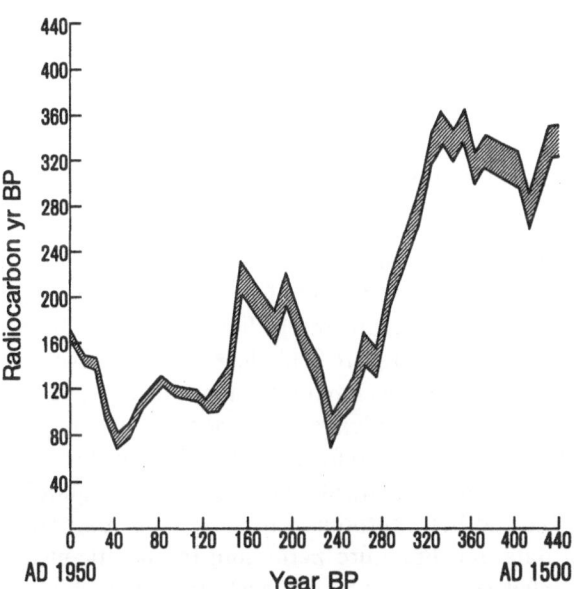

**Fig. 6.** The relationship between $^{14}C$ deduced time scale and tree ring calibrated calendar years. A $^{14}C$ deduced age of 220 years for instance may be compatible with 160, 200, 300 calendar years (BP). Width of curve is twice the standard deviation in the measurements. (Redrawn from Stuiver 1978 [123])

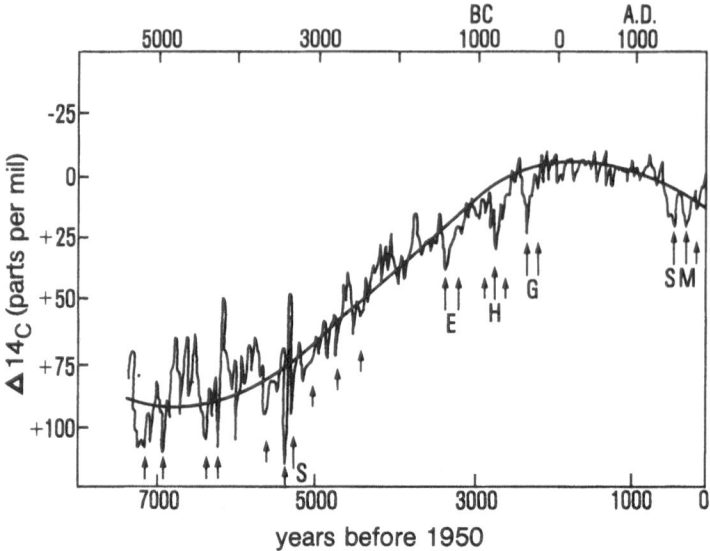

**Fig. 7.** $^{14}$C deviation since 5300 BC derived from analysis of dated tree rings (Damon 1977) with smoothed curve of sinusoidal variation in the Earth's magnetic moment (Lin et al. 1975; adapted from Landscheidt 1983 [67]). Arrows indicate periods of reduced solar activity, which apparently coincide with colder climatic conditions: for instance Maunder and Spoerer minimum with the Little Ice Age (from right: M = Maunder minimum, S = Spoerer minimum, G = Greek minimum, H = Homeric minimum, E = Egyptian minimum, S = Sumerian minimum)

cosmic radiation and sunspot activity and its impact on the atmosphere and the production of radiocarbon. An example is given in Fig. 7.

### Variation of Surface Air Temperature

Weather and climate are the result of a non-uniform distribution of solar energy within the earth-atmosphere system and of the effort to balance this inequality. A primary physical effect of the radiant energy fluxes is the development of a globally and locally differentiated temperature field: the balancing drive. On the other hand the surface air temperature is the climatic element the previous history of which is fairly well documented. By far the longest and most reliable records of this element exist in measured and reconstructed form. This is also true for more representative places all over the earth than for any other element, e.g. precipitation. In addition it is the main variable and best reproduced quantity in climate simulation.

A selection of data and literature pertaining to climatic elements other than temperature—rainfall, sea, lake and river levels, glaciers and ice sheets, windiness etc.—is given for instance by Lamp [60], or is presented in "World Survey of Climatology" edited by Landsberg [65]. The literature is very extensive and rapidly accumulating. The only study to be mentioned here concerns the large-scale changes of precipitation since the middle of the 19th century over the continental areas of the northern hemisphere (Bradley et al. 1987 [8]). This study reveals an increase of more than 20% in mid-latitude (35°–70 °N) precipitation since the middle of this century and is associated with a decrease of temperature. At the same time the lower-latitudes of the northern hemisphere (5° to 35 °N) exhibit a decrease of the precipitation rate of almost the same order. The authors point out that these changes "should be viewed as defining large-scale natural climatic variability". A potential variation of precipitation rate can also be related to the increase of anthropogenic trace gases which absorb radiation, e.g. $CO_2$. However, the different model calculations applied to the problem yield no uniform trend with respect to rain- and snowfall.

A temperature history covering the past million years of the Quaternary era of the northern hemisphere is presented in Fig. 8 covering the entire span of man's presence. A comparison of the temperature variation from the Holocene back to the Riss ice age derived from the Vostok ice core reveals considerable differences in the chronology and also the temperature amplitude. The Vostok data (Fig. 5a) based on a well established isotope–temperature relation and dating standard [54] show a twice as high temperature amplitude of $-11$ °C during the last glacial period and an Eem interglacial about 2 °C warmer than the Holocene compared to the former data compiled in Fig. 8. Neglecting this quantitative differences and considering the logarithmic time scale it should be realized that times as warm as now have been very unusual during this period. With the beginning of recorded human history, $10^4$

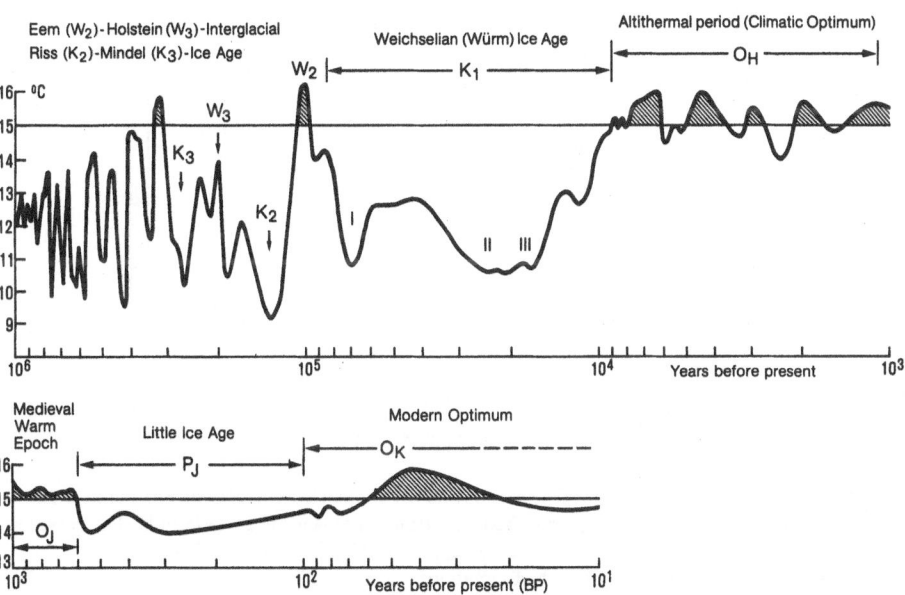

**Fig. 8.** Approximate history of the near-surface temperature of the Northern Hemisphere over the last million years (Quaternary era = Pleistocene + Holocene). $O_H$, $P_J$, $O_J$ are subperiods of the Holocene (postglacial or recent times); they comprise $\sim 1\%$ of the era presented. (After Clark 1982 [14], completed by Schönwiese et al. 1987 [113], modified)

years ago and with the possibility of a higher time resolution, temperature fluctuates with an amplitude of approximately 1 °C. This range of almost 2 °C covers the temperature depression between the climate optima of the Medieval Warm Epoche (A.D. 1000 to 1400) and the midtwentieth century. The difference of 1.5 to 2.0 °C seems to be small compared to the seasonal temperature variations, but

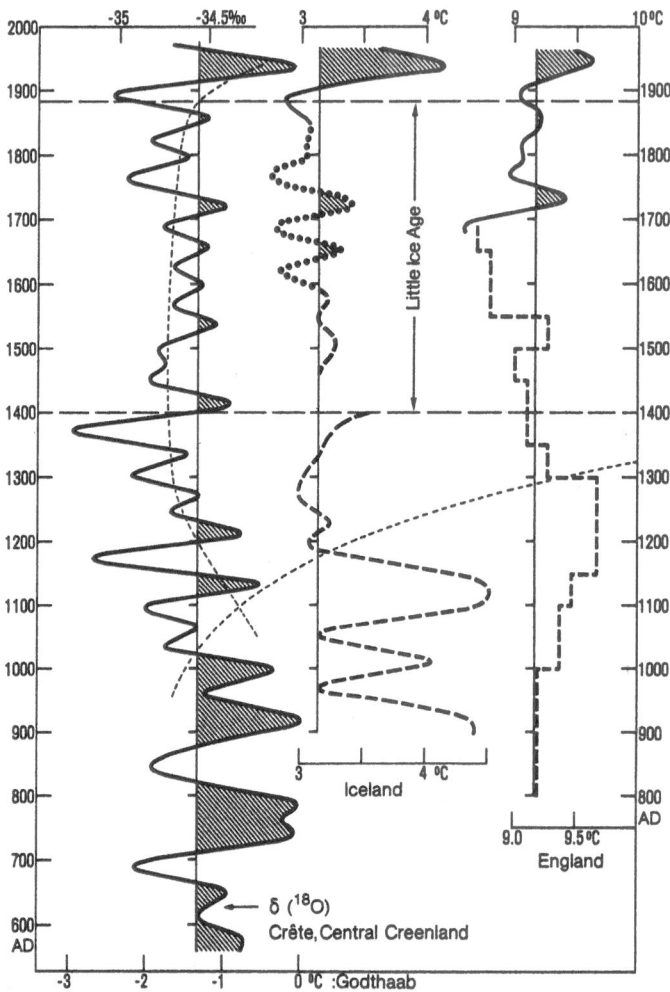

**Fig. 9.** Comparison between $\delta^{18}O$ profile for snow fallen at crête (central Greenland) and near surface air temperature for Iceland and England. (Redrawn from Dansgaard et al. 1975 [19]. The curves are smoothed by a 60-year low-pass digital filter, except for England 800–1700 A.D.)

———————— systematic, direct observations
------------------- indirect evidence
· · · · · · · · · · · · estimated from systematic ice observations
— — — — — onset and duration of "Little Ice Age".

Temperature/$\delta$ relationship taken from Godthaab, West Greenland temperature observations (since 1876)

should not be underestimated. The period from 1400 to 1880 between this two most recent climate peaks exaggerating called "Little Ice Age" is marked by great storms and coastal floods, poor harvests, famine and disease. It is a period of rapid expansion of mountain glaciers in the Alps, Norway, Iceland, Alaska and probably elsewhere with three maxima about 1650, 1750 and 1850, separated by slight withdrawals [140]. Biosphere and cryosphere do not react on short-term but they react on long-term, even minor, variations.

The "Little Ice Age" and its climatic consequences has been documented by the different methods mentioned in the preceding chapter, partly already by instrument record. The downturn of climate can be recognized around A.D. 1100 in Greenland, moving south during the following centuries (cf. Fig. 9). The descent of the timberline by 100 to 200 m in the European sub-alpine mountains reported by Firbas [30] and based on pollen analysis started around A.D. 1200. The decline of the English vineyards, which were successfully cultivated during the high Middle Ages, set in around 1300 with a sequence of remarkably severe winters. And also the northern limits of the vineyards in central Europe moved south [60]. It is not only the cooling, which affected the vegetation, but also the predominantly wet summers and mostly wet springs and autumns. Sometimes, especially in eastern Europe, there seems to have been trouble with heat and drought in the summers.

The onset of the climatic recovery at the end of the nineteenth century is rather abrupt. The change of the average global surface air temperature amounts to 0.5 °C. In both the warming up to around 1940 and the cooling (and wetter [8]) tendency since, the change began to be registered first in the Arctic and in the high latitudes. And it is the northern hemisphere where it was strongest. This is demonstrated in Fig. 10 and in a more differentiated way in Fig. 11. The increase of amplitude is highest in the zone 60 and 80 °N despite its small area and it is highest during the hemispheric winter. The relative climatic optimum around the center of this century

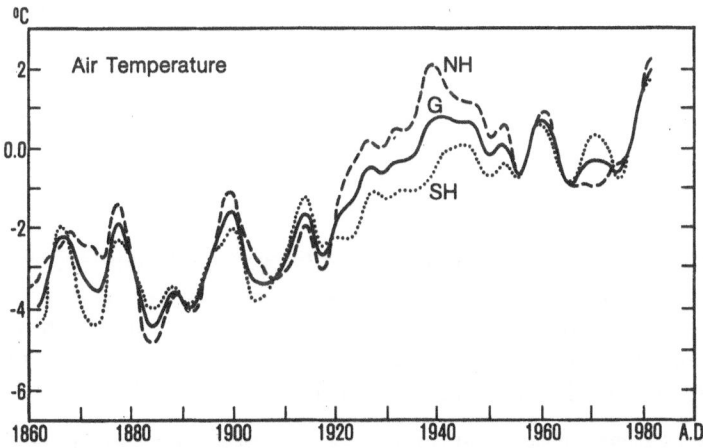

**Fig. 10.** Comparison of northern hemisphere (NH), southern hemisphere (SH) and global (G) near surface mean air temperature variations (without data from the area south of 62.5 °S). Data 10 year low-pass filtered. Departures from the mean for 1951–1970. (Redrawn from Schönwiese 1987 [112])

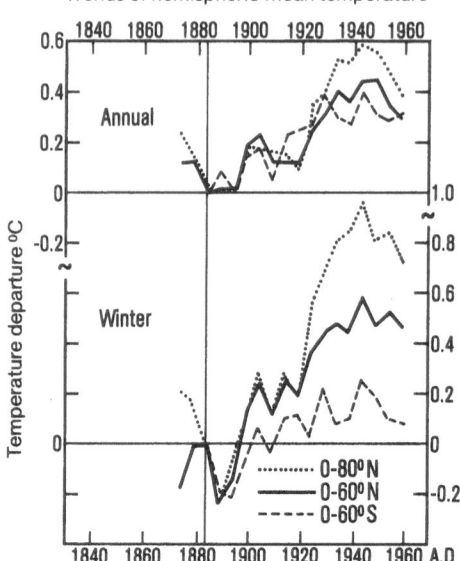

**Fig. 11.** Trends of hemispheric mean temperature from 1870–4 to 1955–9. Successive 5-year means estimated as departures from the mean for 1880–4. (After Mitchell 1963 [82]; adapted from Lamp 1977 [60])

is only weakly developed in the southern hemisphere and it vanishes in the antarctic zone between 60 °S and the south pole. This is demonstrated by instrumental observations and confirmed by isotopic data [2]. The deuterium $\delta$D-curve represented in Fig. 4 indicates an almost continuous cooling of about 2 °C between 1850 and 1980.

The cooling trend observed at least in the northern hemisphere since 1940 seems to have ended in the late 1970s thus stimulating the discussion of the carbon dioxide increase and the greenhouse hypothesis.

## Causes of Climate Variations

To develop a reliable system of modelling or forecasting the future climate, we first have to understand the causes and processes of climatic variations. This understanding comprises the quantification of the elements, which control the key processes and their long-term behavior. Processes which have affected or may affect the climate or, initially, the temperature level are:

— change of solar constant (sunspot blocking and facular emissions)
— global distribution of solar energy (astronomical orbital cycles)
— volcanic activity (planetary albedo)
— change of terrestrial environment (surface albedo)

— change of gaseous composition of the atmosphere (greenhouse effect)
— change of atmospheric particle concentration (cloud cover, radiation transfer, greenhouse effect).

The three processes mentioned first may be subsumed under natural processes. The last three processes however are predominantly controlled by human activities; the known natural process of this kind—the volcanic activity—is specified separately. With respect to the change of the terrestrial environment the classification is correct in a limited sense only. The surface of the earth has been cultivated and so undoubtedly changed by men especially in the middle-latitudes. The present expansion of the deserts, however, the Sahara for instance, cannot or only hypothetically be attributed to human activities.

The waste heat produced by human activities has not been mentioned above. This heat released by industrial plants, household heating, cooling towers etc. occurs near the earth surface and has measurable effects on the micro- to mesoscale climate ($10^2$–$10^8$ m$^2$). An example may be the climatic phenomenon of the city heat island. It is characterized by a temperature increase of 2–3 °C produced by heat fluxes up to $10^3$ Wm$^{-2}$. On macro- or global scales the heat fluxes are considerably smaller, ranging, for the present, around 0.015 Wm$^{-2}$ [132]. Since this energy flux is very small compared with the net radiative energy flux at the earth surface it may be neglected with respect to the global climate. Sensitivity studies to estimate potential climatic effects with increasing waste heat are not known.

**Climatic Impact of Natural Processes**

The solar constant is by far the strongest force, and its variation can directly perturb the surface temperature. Recent measurement, theoretical arguments and climatic measurements suggest the possibility that the solar constant varies significantly on time scales ranging from billions of years (Newman and Rood [86]) to a 11-year sunspot cycle (Hoyt [48]) and even to scales of a few weeks (Willson et al. [143]). But, excluding measurements during the last decade, there is no observational or experimental evidence that the solar constant has been different or significantly smaller in the past [57]. This, however, is indicated by almost all solar models based on basic physical features, which show an increase in the solar luminosity on the order of 30 percent over the last 5 billion years [138]. Yet simple models (see for instance North et al. [89]) of terrestrial climate indicate that a decrease of even a few percent in the solar constant produces a completely glaciated earth. To thaw the ice cover a solar constant even higher than the present value is required. Of course, these models are for the present earth and atmosphere. If the basic physical concepts of stellar evolution are not fundamentally in error, then the result of the solar models is in conflict with the climatic history of the planet earth. This conflict can be avoided, if we presume an earth with continents and oceans in different positions, with substantial contributions of the greenhouse effect from atmospheric components not present in large amounts in the atmosphere now. The biosphere itself may play an important part in changing the atmospheric composition in such a way that the surface temperature of the earth remains nearly constant (quoted in [86]).

The influence of solar activities on the solar constant for at least the last century can be estimated and calculated: the net deficit in irradiance due to cooler sunspots (1000 °C less) and the net surplus due to faculae [49]. Figure 12 shows the monthly mean of the solar irradiance from 1974 through 1981. The percent changes are with reference to a quiet sun irradiance or solar constant of 1369.2 $Wm^{-2}$ based upon measurements on the least-active days in 1980. The calculations have been compared to observations made during 1980. The model results and the observations agree to within $\pm 0.05\%$. Sunspots are the major modulator on this relative short time scale, so the 10.7-year modulation in irradiance is not surprising.

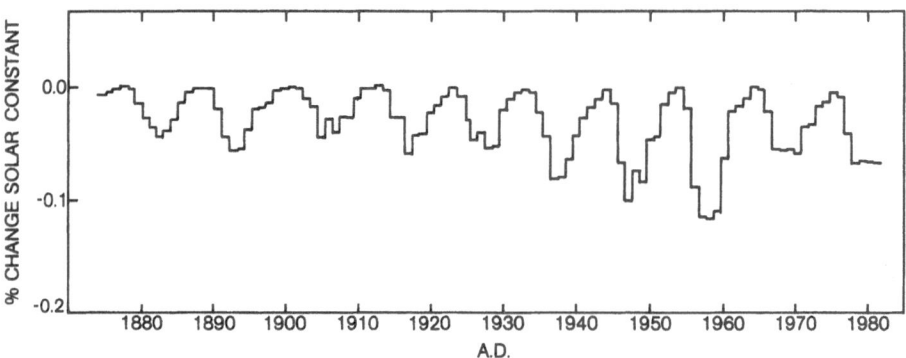

**Fig. 12.** The monthly mean variation in solar irradiance indicated by model calculations for April, 1874, through October, 1981. A quiet sun irradiance of 1369.2 $W m^{-2}$ is assumed (Hoyt and Eddy 1982 [49])

The sunspot activities may be mentioned again, now in another context. This solar activity and its variations are very closely related to ultraviolet radiation mainly to fluctuations of the Lyman-alpha-line ($\lambda = 0.1216\ \mu m$). These fluctuations can only produce negligible changes in the solar constant. But periods of reduced sunspot activities (and reduced $^{14}C$ production)—that means periods of higher solar temperature and solar constant—can be correlated somewhat confusing with times of generally low temperatures prevailing over most of the world. The Maunder (1645 to 1715 A.D.) and Spoerer (1400 to 1510 A.D.) minima for instance coincide with the Little Ice Age (cf. Fig. 7).

Another solar effect, which may be of significance for the solar–climate relationship, has been discussed by Gilliland [38]. Recent studies show periodic variations of the solar radius on an 80 and 11 year time scale. The decrease of solar irradiance with decreasing radius allow—according to Gilliland—a self-consistent explanation of the recent hemispheric temperature trends.

Regular cyclical variations in the earth's orbital arrangements, for instance of the ellipticity of the orbit and of the tilt of the polar axis relative to the plane of the orbit, affect the distance from the sun and the solar radiation available at different latitudes and seasons. These cyclical variations, over periods of about 100,000 (eccentricity), 41,000 (obliquity) as well as 23,000 and 19,000 (precession) years respectively are entirely calculable, and allow us to explain, in outline, the alternations between ice ages and warm interglacial climates, like those of today and historical times [138]. So the glacial-interglacial climatic variations documented by the Vostok ice-core results could be interpreted by an interaction between an orbital forcing "strongly amplified by possibly orbital induced $CO_2$ changes" [36]. In general orbital forcing provides a predictable element of the future. However, it is important to realize that it does not explain the high velocity of some of the climatic shifts and the occurrence of some of the smaller climatic fluctuations such as the temperature amplitude. For these we must look to other causes such as the variations in the sun's output (mentioned above), the change in atmospheric composition [56] or in volcanic activities.

Violent volcanic eruptions such as that of Krakatau in 1883 or Mount St. Helens in 1980 catapult myriads of submicron-size mineral and aerosol particles mainly sulfuric acid droplets and/or ammonium sulfate into the lower stratosphere. The residence time of these particles in the stratosphere ranges from one to two years, or more, depending on the height to which the material is thrown by the eruption. They gradually spread into an increasingly uniform veil which covers the hemisphere concerned or can be observed worldwide. The effect of their partial interception of the solar radiation is an increase of the planetary albedo, that means a loss and interchange of energy for the earth-atmosphere system. These stratospheric aerosol injections and the direct influence and the atmospheric radiation field are the volcanic climate perturbations widely recognized. Theoretical studies show that the stratospheric dust layer warms up while the temperature of the earth's surface and of the lower atmosphere decreases [25]. In the case of the great Tambora Eruption (8 °S, 118 °E) in 1815 a mean temperature reduction in the Northern Hemisphere of 0.4 to 0.7 °C in 1816 has been extracted from observed data—a year without summer [122]. The "nuclear winter" the climatic consequence of a nuclear war—has been simulated and calculated on the basis of a powerful volcanic eruption [95].

Figure 13 represents a synopsis of the surface temperature variation and volcanic activities since 1580. Two different units of measurement for the volcanic activity are used here:

— the "dust veil index DVI" after Lamb [62] which specifies the volcanic dust concentration in the stratosphere of the northern hemisphere
— the "acidity index AI" or $H^+$ ion concentration in the polar ice sheet according to Hammer et al. [42] attributed to volcanic aerosol particles deposited after an eruption.

A detailed statistical analysis of the volcanic impact on climate is given by schönwiese [112]. In general a volcanic forcing of the temperature variation is confirmed. For the last century the correlation becomes even stronger, nevertheless the statistical analysis also indicates that other effects, for instance changes in solar radiation emission or in the gaseous composition of the atmosphere are involved. Besides statistical verification, the study of tree rings offers direct evidence of volcanic events. LaMarche and Hirschboek [59] showed that there is a good

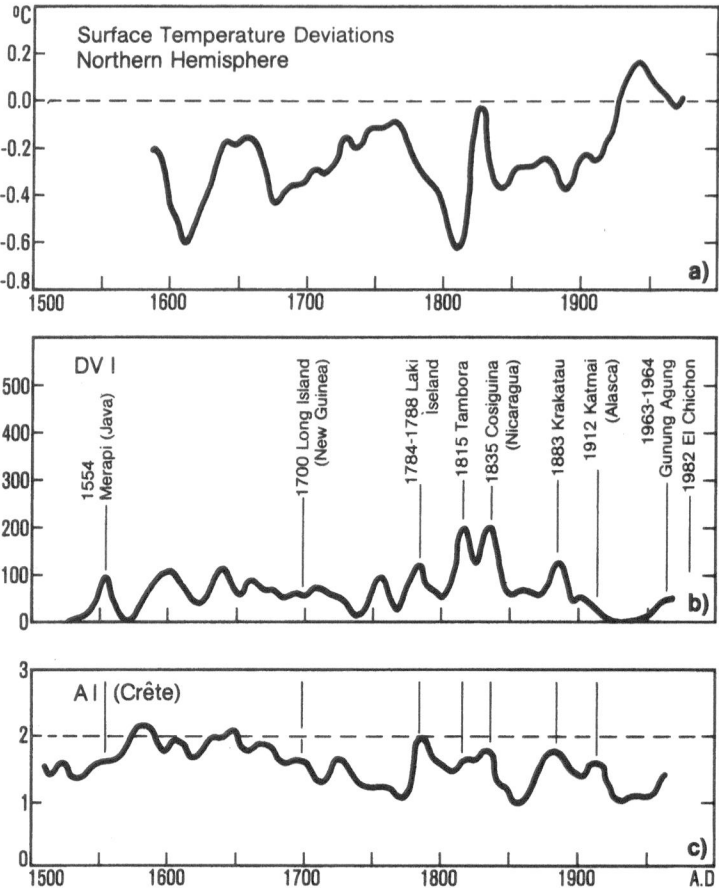

**Fig. 13.** Low-pass filtered time series of climatic data; time periods T < 30 years suppressed. (Redrawn from Schönwiese 1986 [111])

**a)** deviations (anomalies) from 1951–1970 mean surface temperature, northern hemisphere (for marine data 1861–1984, see Jones et al. 1986 [53])

**b)** dust veil index DVI, northern hemisphere
(Lamp 1983 [62])

**c)** acidity index AI, central Greenland, crête ice core measurements
(Hammer et al. 1980 [42])

agreement in the timing of frost events, conserved by annual tree rings, and volcanic eruptions.

The volcanic (and today also the anthropogenic) activities also augment the tropospheric particle and condensation nuclei content of the troposphere. As a consequence we will have more but smaller cloud elements and less rain. Besides the increase of planetary albedo by scattering and reflection of solar radiation back into space caused by the volcanic stratospheric particles, there will be also a further amplification of this effect by an enhanced cloudiness. Clouds have a major effect on the radiative field of the atmosphere. In general the influence of an increasing cloud cover is twofold: on the one hand the increase of the planetary albedo implies a loss of primary solar energy to the earth–atmosphere system, on the other hand, it reduces the outgoing infrared radiation at the top of the atmosphere. Depending on the varying height and distribution, clouds may enhance or reduce the greenhouse effect. The cloud-climate feedback is particulary difficult to determine and a source of uncertainty. There is a lack of long-term cloud data and there are still observational and modelling problems [47, 137].

A further effect of natural processes (for instance glaciation) as well as human activities intruding upon the radiation regime and greenhouse effect is the change in terrestrial environment. As long as the changes by men are performed to cover the basic necessities of life they have to be considered a natural or ecological process. In this sense, the greatest change so far is the clearing of the northern hemisphere temperate forest zone and its conversion to farm land. This may have reduced the surface roughness and so increased the prevailing wind strengths but may not have had a great effect on radiation absorption and heat of the areas concerned. The removal of the tropical rain-forest such as is now occuring in the Amazon basin or contemplated elsewhere increasing the surface albedo and reducing significantly the surface and atmospheric humidity seems to be more serious. Theoretical considerations suggest that the changes in this case will be a reduction of the surface heat absorption and the greenhouse effect, and hence a decrease of convection cloudiness and rainfall, but an increase of surface temperature [132]. The forest decline observed at present in the northern middle-latitudes can be at least hypothetically attributed to climate fluctuations (just as to environmental pollution [92]). A feedback to global climate however is unlikely.

**Climatic Impact of Anthropogenic Trace Gases**

The pertubation in atmospheric trace gases is at present a widely discussed factor in climate change. Trace gases present in the atmosphere influence its thermal structure. Most of these gases have strong absorption bands in the infrared and interact with the IR radiation emitted both by the earth surface and the atmosphere itself and hence contribute to the greenhouse effect. Recent research has been intensified on the man-made increase of $CO_2$ and temporal and spatial variation of $O_3$. Other gases, however, can also affect the temperature structure of the atmosphere. Trace gases can have climatic effects both directly by absorbing infrared radiation and indirectly via chemical reactions with other radiative active gases. Trace gases known or assumed to affect climate are summarized in Table 1. Gases and their products with an atmospheric lifetime of only few days are not specified, for instance $NH_3$, $HNO_3$, $SO_2$. These gases are of local interest only.

A general requirement for interfering effectively in the climate processes is a hemispherical or global distribution of the gas in question. This applies to trace gases which have an atmospheric lifetime comparable or larger than the mixing time of the troposphere. This time period is approximately one year and allows for a uniform mixing. Another but most important condition refers to the spectral properties of the gas. Despite the small amounts of the most trace gases they can

**Table 1.** List of atmospheric trace gases whose variations and inputs have a detectable effect on climate (present situation). For further details and comments see Ramanathan et al. 1985 [100], Visconti 1985 [129], Dickinson et al. 1986 [22]

| Trace gases | Mixing ratio | Lifetime | Increase | Center of main absorption bands | Sources |
|---|---|---|---|---|---|
| $CO_2$ | 347 ppm | 5–10 years | 0.4%/a | 4.5 μm, 14.7 μm | Fossil fuel combustion, decay of organic material |
| CO | 0.12 ppm | 2–6 months | <1%/a* | 4.8 μm | Fossil fuel combustion, oxidation of hydrocarbons |
| $CH_4$ | 1.65 ppm | 4–7 years | 1.5%/a | 6.5 μm, 7.7 μm | Anaerobic decay, combustion of biomass-fossil fuel, natural gas, rice-growing, stock farming |
| $N_2O$ | 0.32 ppm | 20–100 years | 0.25%/a | 4.5 μm, 7.8 μm, 17.0 μm | Anaerobic decay, fertilizer, combustion of biomass-fossil fuel |
| $CCl_4$ | 0.14 ppb | ~50 years | 2.4%/a | 13 μm | Anthropogenic activities (solvent, cleansing agent) |
| $CFCl_3$ (Freon 11) | 0.18 ppb | ~74 years | 5%/a** | 9.2 μm, 11.8 μm | Anthropogenic activities (propellant gas, refrigerant) |
| $CF_2Cl_2$ (Freon 12) | 0.30 ppb | ~111 years | 5%/a** | 8.7 μm, 9.1 μm, 10.9 μm | Anthropogenic activities (propellant gas, refrigerant) |
| $O_3$ (surface) | 30 ppb | 1–3 months | 1%/a | 9.1 μm, 9.6 μm | Natural photochemical process, electrical discharges (indirect: fossil fuel combustion, traffic) |
| $CH_3Cl$ | 0.6 ppb | — | — | 9.9 μm, 13.7 μm | Anthropogenic activities (solvent cleansing agent, propellant gas—prohibited, extraction solvent) |

*Seiler et al., Southern Hemisphere 1984 [114]; **Cunnold et al. 1986 [17] have significant effects on the thermal structure of the atmosphere if they have absorption bands within the 8 to 12 $\mu m$ atmospheric window which transmits most of the thermal radiation from the earth surface and lower atmosphere. In contamination or blanketing of the atmospheric window, a radiatively active gas is able to play a powerful role. Whereas outside the window region an additional gas is only a minor partner dominated by the primary radiative gaseous constituents—water vapor and carbon dioxide [37].

Water vapor is a continuously fluctuating atmospheric constituent. It depends and reacts strongly on the physical conditions and dynamics of the atmosphere and on the physical structure of the surface layer of the earth. Experimental data on the spatial and temporal distribution of water vapor are very uncertain, a long-term variation of the average atmospheric concentration is unknown. Carbon dioxide is, in contrast, well documented [14]. It is homogeneously mixed with the permanent constituents of the air and its mixing ratio increases uniformly. It interacts directly with the infrared or thermal radiation field of the atmosphere. The variation of the atmospheric $CO_2$ causes the most important suspected climatic effect. After doubling the atmospheric $CO_2$ content a surface warming of up to 5 °C may be expected (Table 4) whereas the doubling of other radiatively active gases results in minor surface temperature changes. The estimated positive change for $CH_4$ is in this case 0.2 to 0.4 °C, for $N_2O$ 0.3 to 0.4 °C [132]. Using the concentrations of $CF_2Cl_2$ and $CFCl_3$ listed in Table 1 and considering only the absorption of radiation within the atmospheric window a temperature change of about one tenth of these variations may be expected.

Besides the direct interference in the radiative transfer processes, anthropogenic trace gases may interact through atmospheric chemical processes with the species that affect the temperature distribution. In this respect ozone plays an important role as reaction partner of anthropogenic trace gases ($CH_4$, $N_2O$, CO, $NO_x$, chlorofluorocarbons) which are generally neglected with respect to the effects of carbon dioxide. Ozone is a strong absorber mainly in the near ultra violet range of the solar spectrum (Hartley and Huggins bands) determining the thermal structure of the stratosphere; in the visible range of the spectrum (Chappuis bands) it heats the strato- and troposphere and the absorption and emission of infrared radiation predominately in the atmospheric window region ($\lambda = 9.6 \mu m$) sustains the greenhouse warming effect. The observed [87] and expected reduction in total ozone content and the change in vertical distribution—depletion in the stratosphere, increase in the troposphere—probably will cause an appreciable cooling of the middle and upper stratosphere and a slight warming of the lower stratosphere and of the troposphere [9, 130]. The most serious consequence of a change in $O_3$ however will be the impact on the radiation shield of the earth and the increase of ultra violet radiation ($\lambda < 300$ nm) at the earth surface with damaging effects to biological systems. A reduction of total ozone of about 10% for instance will cause an increase of skin diseases for white people between 25% and 130%, depending on the nature of the disease and the geographical latitude [27]. The chemical, photochemical and dynamical atmospheric processes said to be responsible for the variation of the atmospheric ozone distribution are very complex [9, 16]. A detailed display is beyond the scope of this article.

## Climatic Impact of Anthropogenic Aerosol Particles

Atmospheric particles are either injected directly into the atmosphere or produced by gas to particle conversion. The predominate mass of the global atmospheric particle material is of natural origin (Table 2). An estimated fraction of 9 to 12% has to be attributed to anthropogenic activities. During the residence time of the particles within the atmosphere they agglomerate by the coagulation processes and they are well mixed. The life time of an individual particle depends on its size. For the smallest particles with a radius of approx. $r = 10^{-3}$ $\mu$m, the life time is limited by the coagulation process and is in the order of seconds to minutes. To tropospheric particles in the optical active size range of $r \approx (3 \times 10^{-2}$ to $3 \times 10^{0})$ $\mu$m an average life time of a week up to a month can be assigned. The life and residence time of stratospheric particles of this size range—predominately of volcanic origin—is of the order of years and for that reason detectably influencing climate. The residence times of the larger particles are increasingly governed by sedimentation and drops down to 10 seconds for the upper limiting size of $r = 10^{3}$ $\mu$m. The tropospheric removal process for the particles within the optical active size range is rainout; the central physical process is condensation and specially the more soluble particles are used as condensation nuclei. This favours a crude fractionation: the more insoluble particles—for instance anthropogenic combustion products, in particular the strongly absorbing soot particles—are enriched in the troposphere, intensifying the specific interrelation between atmospheric particles and climate. Changing optical

**Table 2.** Estimates of global atmospheric particle mass, in Mt/a, produced from natural and anthropogenic sources (after Jaenicke 1980 [52], supplemented)

| | | |
|---|---|---|
| *Natural* | | |
| Direct particle production | | |
|   Sea spray residue | | 1000–2000 |
|   Windblown mineral dust | | 60–360 |
|   Volcanic emissions | | 4 |
|   Meteoric debris | | 1–10* |
|   Biogenic materials (spores, pollen etc.) | | 80 |
| Particles formed from gases (converted | | |
|   sulfates, nitrates, hydrocarbons) | | 1319 |
| | subtotal | 2464–3773 |
| | | |
| *Man-Made* | | |
| Direct particle production (e.g. soot, | | |
|   smoke, road dust, etc.) | | 54–126 |
| Particles from the conversion of | | |
|   anthropogenic gases | | 270 |
| | subtotal | 324–396 |
| | total | 2788–4169 |
| Man-made fraction of total particle mass | | 9–12% |

* Rosen, J. M.: Space Sci. Rev. 9, 58–89 (1969)

aerosol properties modulate the atmospheric radiation field and determine the scattering losses and absorption profiles of radiative energy of the atmosphere and consequently influence the climate. Besides particle size distribution and number concentration the absorption index of the particle material is the most significant quantity. All these "optical" quantities are influenced by anthropogenic activities.

A long-term observation of the atmospheric aerosol content comparable to the observation of the global $CO_2$ background concentration is not known. At present a reliable assessment of the trends in aerosol characteristics—past and future— seems to be impossible. The temporal and spatial distribution is very variable and the relative short residence time of the tropospheric particles of the order of weeks prevents a homogeneous global mixing [99]. The anthropogenic particle sources are predominately located in the northern middle-latitudes. In these regions the most distinct climatic effects should be expected. An influence of the anthropogenic, atmospheric aerosol particles on the recent climate, such as in the case of stratospheric volcanic dust [112], can not yet be proven.

The climatic effects initiated by an enhanced particle load of the atmosphere may be evaluated by a doubling of the actual particle concentration (Fig. 14). The trend

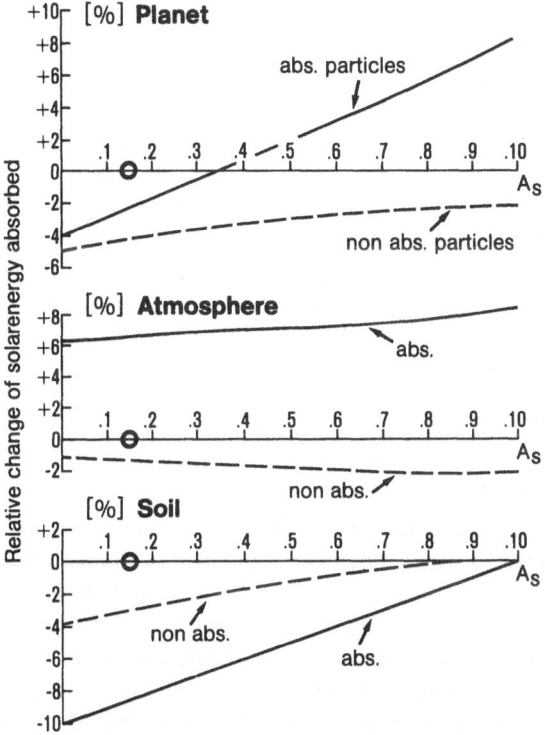

**Fig. 14.** Relative variation of solar energy absorbed by the planet earth caused by a doubling of the actual number concentration of aerosol particles. Assumptions: cloud cover 50%, albedo of clouds 46%, total atmospheric water vapor 2.28 cm ppm, mean absorption index of the actual and added absorbing particles $\bar{k} = 0.02$, $\bar{n} = 1.5$ real refraction index (Eiden 1979 [25])

and the deviation from the actual situation is given as a function of the surface albedo $A_s$ (the present average global value amounts to $A_s \approx 15\%$). In the case of additional nonabsorbing and strongly absorbing particles, the solar (visible) radiative energy reaching the earth surface will be reduced. Consequently the solar energy absorbed by the surface layer will be diminished; the reduction of solar energy is highest with absorbing particles. In any case, the consequence will be a cooling of the surface layer. As long as the nonabsorbing particles predominate the atmosphere will also lose energy and cool down; it will gain energy only with absorbing supplementary particles, for instance soot, and warm up. With an excess of nonabsorbing particles the whole planetary system loses energy in any case: the back scattering of solar energy into space increases. That is also true for an absorbing aerosol as long as the surface albedo is low. With increasing albedo for instance above a snow field with an additional absorbing particle load the planetary system will acquire energy. A high surface albedo prolongs the optical path of solar radiation within the atmosphere and so the absorption probability. In any case, whether the planet loses or gains energy, the consequence will be a redistributional absorbed radiation energy within the system: an excess of absorbing particles will favour and heat the atmosphere and will cool down the earth's surface layer.

The final conditions of the system atmosphere—the surface layer can only be determined considering all the relevant feedback mechanisms of the system. That means more detailed knowledge about variation of the physical properties of the atmospheric particle population. An excessive increase of the small-sized particle concentration ($r < 1 \ \mu m$) essentially consisting of strong-absorbing substances cause a heating of the atmosphere by absorption of solar energy. It screens the infrared emission from the surface and favours, like an absorbing gas, the atmospheric greenhouse effect. An excessive increase of the coarse particle fraction ($r > 1 \ \mu m$) consisting mainly of weak absorbing mineral constituents [26] favours an anti-greenhouse effect [58]. This is due to scattering losses in the solar spectral region and low absorption and emission in the infrared. This anti-greenhouse effect caused by coarse and low absorbing particles of all size compensates and masks or even overcompensates the greenhouse effect by the increasing concentration of absorbing gases. Both, the increase of a low absorbing atmospheric particle load and a simultaneous decrease in the concentration of the absorbing gases (e.g. $CO_2$), enhance the cooling of the atmosphere as indicated by the Vostok ice core measurements (Fig. 5).

Another important physical feature of atmospheric particles is the spectral dependence of the absorption index of the particle substance. In the solar region of the spectrum (Fig. 15) the absorption index $\bar{k}$ of the bulk material shows an almost continuous increase with wavelength $\lambda$. This means that the absorption coefficient $\bar{K} = 4\pi\bar{k}\lambda^{-1}$ and hence the absorption of radiation show no spectral selectivity in the region of the solar spectrum. The weak band structure for $\lambda > 1 \ \mu m$ can be attributed to liquid water incorporated in the particles. According to the optical properties the particles may be grouped into two classes. Particles from highly polluted areas yield significant higher absorption indices than those from clean air regions. The higher absorption can be explained by an admixture of carbonaceous residues from combustion processes or—as in the case of the Tsumeb samples by an

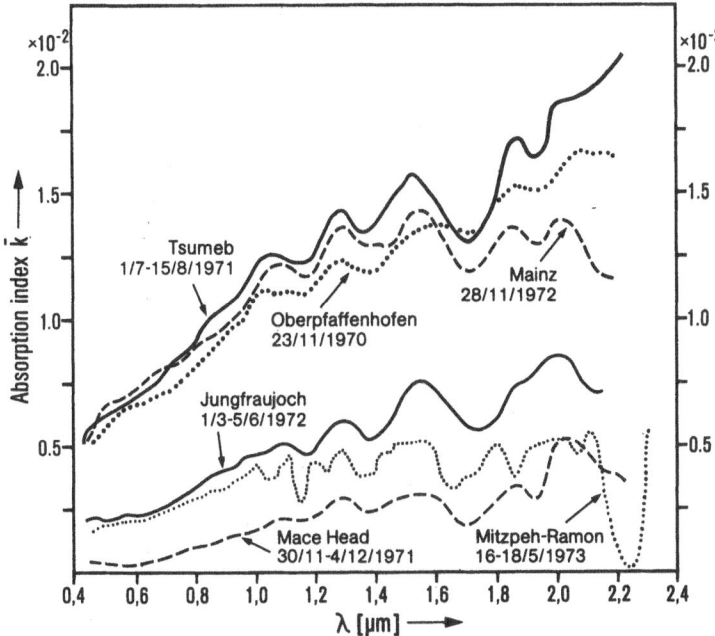

**Fig. 15.** Bulk absorption index $\bar{k}$ of atmospheric aerosol particle samples versus wavelength $\lambda$ (Ebert monochromator), solar spectral region. Samples taken at Mainz and Oberpfaffenhofen (FRG, polluted atmosphere), Mace Head (west coast of Ireland, marine atmosphere), Jungfraujoch (Switzerland, continental atmosphere—background aerosol), Tsumeb (South-Africa, continental atmosphere—dry plant abrasive), Mitzpeh–Ramon (Israel, continental atmosphere—mineral dust). Bulk density of particle material assumed: $\bar{\rho} = 2\ \text{g cm}^{-3}$. (Adapted from Eiden et al. 1987 [26])

admixture of carbonaceous abrasion dust from dry plant material. Very low absorption values are found for maritime aerosol particles (Mace Head/west coast of Ireland).

The spectral dependence of the bulk absorption coefficient is different in the terrestrial or infrared region of the spectrum (Fig. 16). With the exception of the maritime sample the curves show a distinct band structure. A clear identification of the bands and a correlation to definite molecules however is difficult. The broad band with the spectral center at $\lambda \approx 9.2\ \mu\text{m}$ and the band at $\lambda \approx 16.3\ \mu\text{m}$ can be assigned to the $SO_4^{2-}$ ion. The weak bands at $\gamma \approx 3.2\ \mu\text{m}$ and $\gamma \approx 7.1\ \mu\text{m}$ belong to the $NH_4^+$ ion. The maritime samples taken at the west coast of Ireland during west wind influence do not show this marked band structure. The dominating constituents of this samples—NaCl and KCl—have no bands in the spectral range analyzed. Samples taken at this coastal station during east wind episodes, i.e. from the island, again have the characteristic $SO_4^{2-}$ and $NH_4^+$ bands. Silicates, quartz or quartzous minerals, which can always be found at least in continental particle samples, broaden the 9.2 $\mu$m-band. The silicate shoulder ($\lambda \approx 8.6$–8.8 $\mu$m) can clearly be noticed in the particle sample of Mitzpeh–Ramon taken during a sandstorm. A climatic response of the atmospheric aerosol in the infrared spectral range can be expected mainly by the strong band absorption around $\lambda = 9.2\ \mu m$, i.e. in the atmospheric window region ($8\ \mu m \leq \lambda \leq 12\ \mu m$). The particle influence outside the window region is obscured by the strong radiative effects of water vapor and carbon dioxide.

Another important climatic factor already mentioned in the first section of this chapter is cloud cover. An increasing number of soluble particles raises the number of condensation nuclei and accordingly the number of cloud droplets. Assuming an unchanged water vapor supply, this rise of cloud droplets may reduces the formation of rain drops and increases the lifetime of clouds. The same holds for an "overseeding" of clouds with man made ice nuclei. In both cases a further growth of droplets or ice crystals is inhibited by competition for water vapor between the myriad of cloud elements. But there are also studies reporting a triggering role of anthropogenic particles in increasing precipitation and decreasing cloud cover. In any case a growing cloudiness raises the planetary albedo and reduces the solar energy absorbed by the planet. On the other hand the presence of clouds lowers the temperature of the radiative surface in the atmospheric window and correspondingly the outgoing radiation, leading to a positive greenhouse effect. At present it is difficult to foresee the scale and the sign and of the greenhouse effect due to this factor [58].

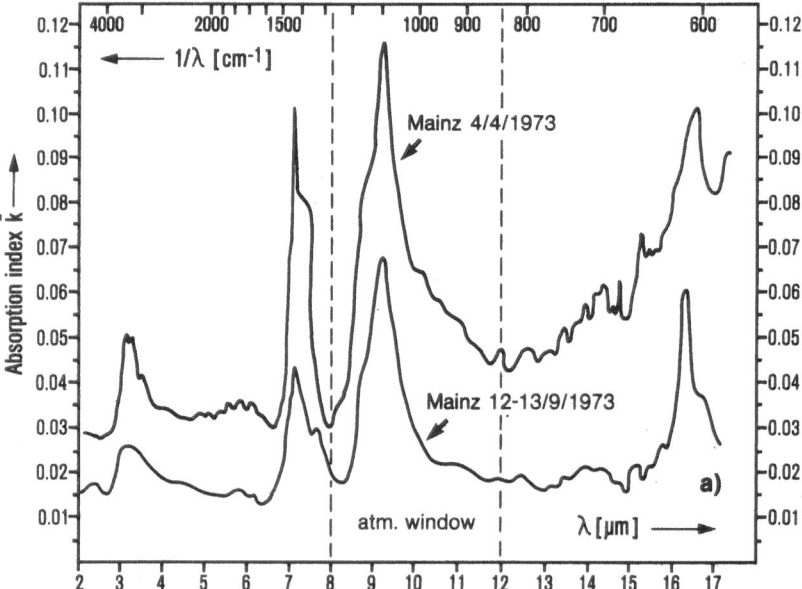

Fig. 16(a)

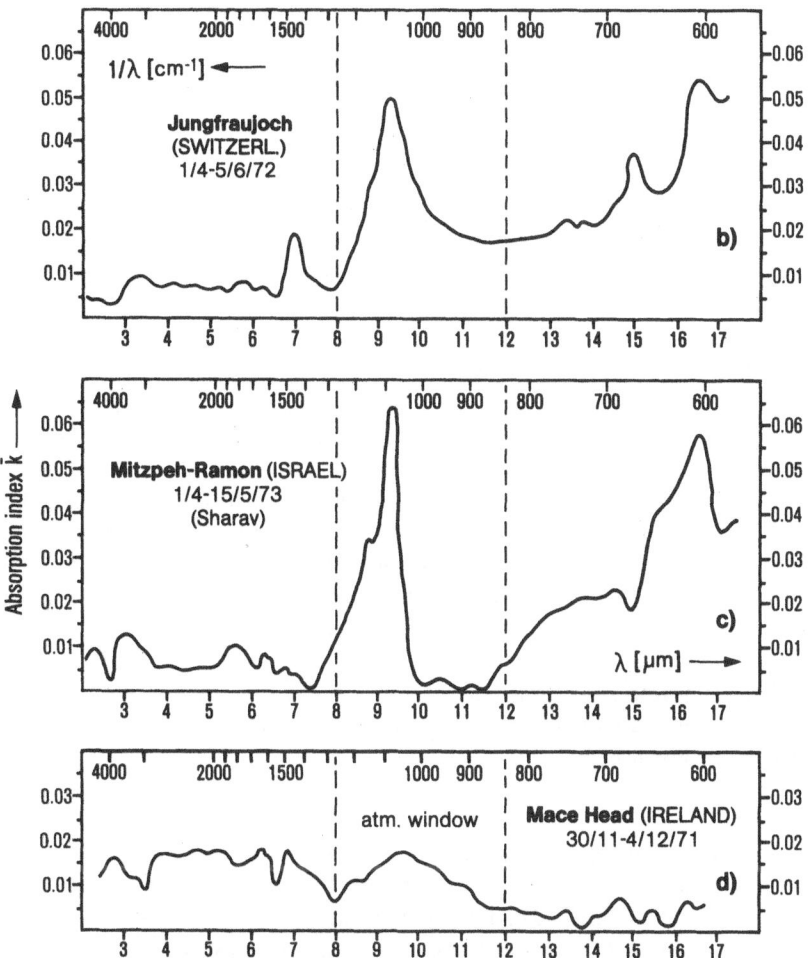

**Fig. 16.** Bulk absorption index $\bar{k}$ of atmospheric aerosol particle samples versus wavelength $\lambda$ (Ebert monochromator), terrestrial or infrared spectral range. Samples taken at
**a)** Mainz (FRG)                        polluted atmosphere
**b)** Jungfraujoch (Switzerland)          continental background aerosol
**c)** Mitzpeh–Ramon (Israel)              continental mineral aerosol
**d)** Mace Head (West coast of Ireland) maritime aerosol Bulk density of particle material assumed:
$\bar{\rho} = 2\,\mathrm{g\,cm}^{-3}$.
(Adapted from Eiden et al. 1987 [26])


## Climate Modelling

The modelling of the global climate has different, more or less coupled objectives. One principal goal concerns the broad changes of the overall climatic system as a result of the changes in its external and internal parameters. It comprises all attempts to find reasons for all these changes. Based on the knowledge acquired from the past, models are used to simulate the present climate. But the most pressing development at present concerns climate prediction. Beyond the natural changes and fluctuations man is trying to estimate the consequences of his own activities on local and, ultimately, on the global climate [55].

Several types of mathematical climate model have been developed. They differ in the comprehensiveness of their treatment of the components of the climatic system and in the physical and mathematical disposition.

Thermodynamic models try to predict temperature or temperature distributions but they neglect the motion field or treat it only in a crude simplified way. There are two main classes of this model: the energy balance models (EBM) and the radiative convective models (RCM).

The EBMs are models with only horizontal dimensions, i.e. the earth surface coordinates. The column of the earth-atmosphere system assigned to a distinct surface element is characterized by only one number, for instance the sea level temperature. By zonal averaging the models are reduced to one-dimensional climate models. The development of this type has been stimulated by Bodyko [11] and Sellers [116]. It is a simple but nevertheless powerful research tool. The adequate basic assumption in the latitude-dependent models is that the energy which enters each latitude belt is exactly balanced by the energy lost during the year: solar energy absorbed = infrared radiation energy emitted + net horizontal energy transported out.

In their simplest zero-dimensional form the EBM determines the effective radiating temperature of the planet $T_p$. In this case the condition of the radiation equilibrium is given by

$$4\pi R^2 \sigma_p T_{p^4} = S_0(1 - A_p)\pi R^2$$

where R is the radius of the planet; $S_0$ is the solar constant [138]; $\sigma_p$ is a planetry emission factor (in the case of a blackbody, $\sigma_p$ is the Stefan–Boltzman constant); and $A_p$ is the planetary albedo [102]. If the planet earth is assumed to emit radiation as a blackbody, balancing the solar radiation absorbed, then, using $A_p = 0.3$, the effective temperature of the whole planet, would be $T_p = 254.5$ K ($= -18.6\,°C$). This is a considerably lower temperature than the observed global mean surface temperature $T_s = 287.35$ K ($= 14.2\,°C$). This higher surface temperature is caused by the infrared-absorbing gases, aerosol particles and cloud elements, in total representing the greenhouse effect. The emission of the earth surface is thus lowered by a factor of $\varepsilon_p \approx 0.61$ depending on the composition of the atmosphere. The magnitude and change of this factor $\varepsilon_p$ can be determined by EBMs.

A fundamental parameter is the sensitivity $\beta_0$ a quantity to measure the change in global average temperature due to a 1% change in solar constant. The sensitivity parameter is defined by

$$\beta_0 = \frac{S_0}{100} \cdot \frac{d\,T_0}{d\,S_0}.$$

The subscript refers to the global average values. According to North et al. [89] $\beta_0$ "is the first quantity to compute because the sensitivity of the model to any perturbation is roughly proportional to $\beta_0$".

Assuming a black planet the simple zero-dimensional model traced above gives

$$\beta_o = \frac{T_p}{400} = 0.63 \text{ K}$$

It is a standard for comparison with all climate models and represents the sensitivity of a system with no feedbacks. The feedback processes effect the radiation fluxes absorbed and emitted and they effect the sensitivity of the system. Surface albedo, cloudiness, aerosol particles, gaseous composition of the atmosphere etc. are elements of the atmospheric feedback system, which may be introduced into the model by treating the physical processes in a parameterized form. (The character-istic size of the physical process is generally smaller than the smallest size resolved by the model.) A review about the energy balance models is given by North et al. [89].

The radiative–convective model (RCMs) computes the equilibrium vertical temperature profile of an atmospheric column and the temperature. Altitude is the only dimension considered in this kind of model. The atmospheric composition, the surface albedo and the incident solar radiation are given. The model includes the transfer process of solar and infrared (terrestrial) radiation, the turbulent heat transfer between the earth surface and the atmosphere and as an essential characteristic dry and moist convection, clouds and the vertical distribution of released latent heat.

The RCMs do not give any information about regional and latitudinal tempera-ture distributions. They do offer reasonable first estimates of the sensitivity of the global surface temperature to changes in radiatively active gases. The global surface temperature changes predicted by the RCM are in good agreement with those obtained from the more complex three-dimensional general circulation models. But the RCM-result is achieved with considerably less computer space and time. This is also true for the calculation of detailed radiative processes, thereby presenting valuable information on the relative importance of such processes. Beyond the question of computer resources the radiative–convective model allows us to deduce cause–effect relationships.

It is much more difficult to analyze physical processes and to test their relevance in a general circulation model (GCM) [101]. The GCMs are based on a fundamen-tal set of nonlinear partial differential equations compiled in Table 3. Considering the different properties of the atmosphere and of the oceans they describe the state of both of these two fluid systems. The solution of the problem for appropriate, realistic boundary or surface conditions (e.g. [115]) can only be obtained by numerical mathematical methods. The GCM generally allows for the prediction of the day-to-day evolution of large-scale wether systems and longer-term climate processes. Since the calculations are very time-consuming and cumbersome, climate studies concerning only special questions or tendencies are based for instance on fixed present day ocean and land surface temperatures. For a comprehensive modelling of longer-term climatic changes the two-way interaction between atmosphere and ocean has to be taken into account. The feedback of atmospheric and oceanic circulations has been considered for instance in the exacting climate studies of Manabe et al. [72], Washington et al. [135], Pollard [97] and Nihoul

**Table 3.** Fundamental equations describing the general circulation and the state of the terrestrial atmosphere

---

$$\frac{d\vec{v}}{dt} = -2\vec{\Omega} \times \vec{v} - \rho^{-1}\Delta p + \vec{g} + \vec{F}$$

Conservation of momentum
(Navier–Stokes equation;
Newton's second law of motion)

$$\frac{d\rho}{dt} = -\rho\nabla \times \vec{v} + G - D$$

Conservation of mass
(continuity equation)

$$\frac{dU}{dt} = -p\frac{d\rho^{-1}}{dt} + H$$

Conservation of energy
(first law of thermodynamics)

$$p = \rho R_a T$$

Ideal gas law
(idealized equation of state)

---

Symbols

t  time
$\vec{v}$  velocity rel. to rotating earth
$\vec{\Omega}$  angular rotation vector of the earth
$\vec{g}$  force of apparent gravity per unit mass
$\vec{F}$  frictional force per unit mass
p  atmospheric pressure
$\rho$  density of air (water vapour included)
T  atmospheric temperature
$R_a$  specific gas constant of air
U  internal energy per unit mass $(= c_v T)$
$c_v$  specific heat at constant volume
H  heating rate per unit mass (including heat of condensation and radiational heating)
G  rate of nascent atmospheric (gaseous) mass per unit volume (describing evaporation)
D  rate of disappearing atmospheric (gaseous) mass per unit volume (describing condensation)

[88]. It is beyond the scope of this article to provide a comprehensive catalogue of the numerous different climate model arrangements; only the more general studies by Hasselmann [46], Simmons and Bengtsson [118], Hansen et al. [45], Schlesinger [106], as also Berger and Nicolis [5] may be additionally mentioned here.

The increase in carbon dioxide is the best documented change of one of the atmospheric gaseous constituents and research has therefore been concentrated on modelling $CO_2$-induced climate changes. All model estimates show an increase of surface temperature [14, 107] and as Wigley pointed out: "although there are a few dissenters, it is now believed likely that increasing atmospheric $CO_2$ concentration may well cause the globe to warm substantially over the coming decades" [139]. Controversial objections by Idso [50, 51] have been clarified in detail by Potter et al. [98].

The standard reference for comparing alternative models is the globally and annually averaged surface temperature $T_s$ due to doubled $CO_2$: 300 ppm to

**Table 4.** Calculated global-mean air temperature change due to doubling $CO_2$ (300 to 600 ppm) in climate models since 1980 (after Schönwiese 1986 [110], supplemented)

| Authors and year of publication | | Estimated temperature change |
|---|---|---|
| Manabe and Stouffer (1980) | [73] | 2.0 K |
| Manabe and Wetherald (1980) | [74] | 3.0 K |
| Gates, Cook and Schlesinger (1981) | [35] | 0.2 K |
| Manabe, Wetherald and Stouffer (1981) | [75] | 2.4 K[a] |
| Wetherald and Manabe (1981) | [136] | 2.4 K |
| Cess and Goldberg (1981) | [12] | 1.5–1.8 K |
| Hansen et al. (1981)—FTT | [43] | 2.8 K |
| Hansen et al. (1981)—FTA | [43] | 1.4 K |
| Schlesinger (1983) | [105] | 2.0 K |
| Washington and Meehl (1983) | [133] | 1.3 K |
| Alexandrov et al. (1983) | [1] | 1.4 K |
| Mitchell (1983) | [80] | 2.9 K |
| Washington and Meehl (1984) | [134] | 3.5 K |
| Hansen et al. (1984) | [44] | 4.2–4.8 K |
| Spelman and Manabe (1984) | [120] | 2.8 K[a] |
| Mitchell and Lupton (1984) | [81] | 2.9 K[a] |
| Peng et al. (step funct. doubling) (1987) | [91] | 2.6 K |
| Peng et al. (transient doubling) (1987) | [91] | 1.5–2.0 K |
| Dickinson et al. (1987) | [23] | 2.2 K |
| Tricot and Berger (1987) | [125] | 0.3–1.4 K |

[a] = Half of $4 \times CO_2$ result
FTT = Fixed cloud top temperature
FTA = Fixed cloud top altitude

600 ppm of $CO_2$ (Table 4). The results by both, simple and more sophisticated models, depend largely on the adopted physical scenario, the parameterization and reactive couplings considered. A study by Schlesinger [107] shows that surface energy balance models produce a higher scatter of the estimations of the global mean temperature ($\Delta T_s$: 0.2 to $\sim 10$ K) than the radiative–convective models ($\Delta T_s$: 0.5 to $\sim 4.5$ K). The physically more realistic three-dimensional general circulation models tend to predict the smallest temperature variations ($\Delta T_s$: 1.3 to $\sim 4.5$ K). They more efficiently capture feedback and transient processes, hydrologic cycle and cloud formation, the circulation and thermal diffusivity of the oceans, the dependance of snow and ice cover on these processes.

The pre-industrial $CO_2$ value of $270 \pm 10$ ppm (around 1850 A.D.) may be doubled in the second half of the next century (Fig. 17). However, as shown by Wigley [139] and Wang et al. [131], the existence of the other anthropogenic trace gases (Table 1) may substantially accelerate this effect. These trace gases, including ozone and water vapor, primarily affect the solar and infrared radiative transfer and energy exchange of the atmosphere. In general, the radiation processes—scattering, absorption and emission—can be calculated with some accuracy (Bolle [7]). The model economy and special model set up, however, require a high level of

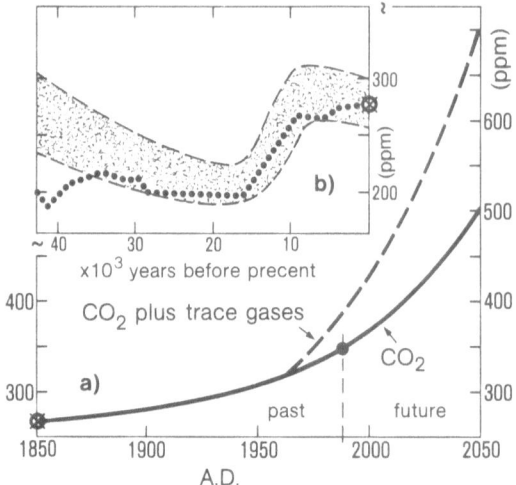

**Fig. 17. a)** Past and estimated future mean yearly atmospheric $CO_2$ and the $CO_2$ equivalent trace gases concentrations. (Wigley 1984 [139])
**b)** Generalized estimate of atmospheric $CO_2$ during the last 40,000 years, deduced from Greenland and Antarctic isotopic ice core measurements (Neftel et al. 1982 [85]). The sign $\otimes$ marks the preindustrial $CO_2$ concentration (1985 [100]). The dotted line represents more recent antarctic measurements [4] (compare Fig. 5)

approximation or parameterization associated with accuracy and speed. The development of a suitable parameterization scheme for calculating radiative heating or cooling caused by the trace gases, including cloud and aerosol particles, is therefore a central, constantly new task fraught with problems [121]. The introduction of the hydrologic cycle with cloud generation and the formation of the different cloud types is still more complicated. The same is true for the global consideration of the atmospheric aerosol; the particle size distribution and the physical particle properties, which have to be taken into account, show a geographical dependence. Nevertheless the influence of the atmospheric aerosol cannot be neglected. As recently discussed by Grassl [39], the change of the net radiation flux caused by an increase of the particle load of the atmosphere is of the same order and opposite to the change caused by the increase of $CO_2$; the $CO_2$ and trace gas greenhouse effect may thus be counter-balanced. This estimation has been based on the present particle emission with a 20% soot fraction and a doubling of $CO_2$.

Besides the direct radiative effects of anthropogenic trace gases the one-dimensional coupled photochemical climate model (RCM) of Brühl et al. (1988 [9]) takes into account the potential interactive chemical and photochemical processes too. The model estimates, in particular, the greenhouse effect generated by the anthropogenic trace gases with a long lifetime ($CH_4$, $N_2O$, $NO_x$, CO, $O_3$ and the chlorofluorcarbons $CF_2Cl_2$ and $CFCl_3$): the study is centered on ozone. It considers feedbacks between temperature, radiation and the chemical processes, as well as the heat transfer to the oceans. Cloud feedbacks are disregarded. The resulting temperature profiles and chemical composition of the atmosphere are hemispherical averages. The increase of the surface temperature calculated by the model almost equals the increase caused by $CO_2$. From the onset of the industrial era ($\sim$ 1860 A.D.) till now an increase of about 0.6 °C has been calculated and a further rise of about 1.2 °C by 2050 A.D. The potential change of the vertical ozone and temperature distribution since 1860 is represented in Fig. 18. The model does not explain the rapid decrease of stratospheric ozone observed, i.e. the occurrence of the antarctic "ozone hole"; the reason may be that the atmospheric circulation is not taken into account.

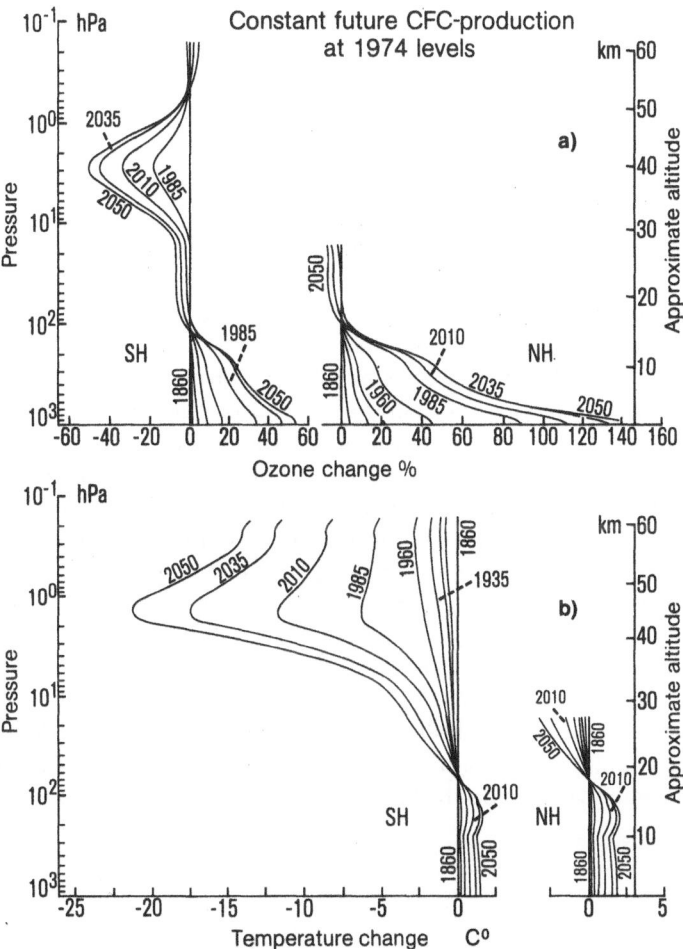

Fig. 18. Vertical changes in a) ozone concentration and b) temperature of the northern (NH) and southern (SH) hemisphere starting 1860 A.D. up to 2050 A.D. The computations are based on the assumption of a continued production of the chlorofluorocarbons (F11 and F12) at the rates of 1974, the year with maximum emission. The calculated stratospheric profiles show almost no differences between the hemispheres. The different natural and anthropogenic emissions of trace gases in the two hemispheres are taken into account. (Adapted from Brühl and Crutzen 1988 [9])

During recent years, well-documented synopses of Northern Hemisphere temperature records extending back to 1881 have been published. These observational data, however, do not prove unambiguously the $CO_2$ induced temperature increase or "greenhouse" effect postulated by the model calculations. Schönwiese [110] showed that the $CO_2$-effect can be "conditionally" verified by statistical multiple regression technique. A principal difficulty is that changes in climate also occur naturally from other, different causes. Variations of solar constant, volcanic aerosols, mineral dust etc. mask or compensate $CO_2$ induced warming. Moreover, it is possible that the thermal inertia of the oceans causes the actual atmospheric temperature increase to lag several decades behind the equilibrium temperature for any change of atmospheric $CO_2$ concentration [21, 107].

## Conclusion

This chapter deals with the impact of atmospheric composition changes on the earth's energy budget and climate represented by temperature. All studies which consider the variation of one or more atmospheric constituents and which are based on known physical relations demonstrate that the observed changes of the physical and chemical composition of the atmosphere also yield significant climatic changes. The increase of infrared absorbing trace gases enhances the atmospheric greenhouse effect and thus the surface air temperature. The increase of absorbing aerosol particles lower the energy consumption of the planet earth by cloud forming processes, the absorption and reflection of electromagnetic radiation. This implies a reduction of the surface temperature. Reliable quantitative estimates of all these effects are not possible because of the lack of sufficient data on the concentration of the trace gases and on the geographical distribution of the atmospheric aerosol particles, as well as incomplete knowledge of the significance of the different feedbacks within the climate system and their effectiveness. Another class of problems consists of the potential influence of natural external forces, such as change in solar constant and astronomical conditions. Only the complete understanding of all the climate processes and of their significance and the exact "weight" of the different feedbacks will allow an adequate description of the past and a reliable predictability of the future atmospheric conditions exceeding a mere trend prognosis. Some scientists doubt that this can ever be achieved [119].

## References

1. Alexandrov V V, Arkhipov P L, Parkhomenko V P, Stenchikov G L (1983) Izv. Akad. Nauk SSSR, Fiz. Atmos. Okeana 19: 451
2. Aristarain A J, Jouzel J, Pourchet M (1986) Clim. Change 8: 69
3. Barber K E (1982) Peat-bog stratigraphy as a proxy climate record. In: Harding A F (ed) Climatic change in later prehistory. University Press, Edinburgh
4. Barnola J M, Raynaud D, Korotkevich Y S, Lorius C (1987) Nature 329: 408

5. Berger A L, Nicolis C (eds) (1984) New perspectives in climate modelling. Elsevier, Amsterdam
6. Beug H J (1982) Vegetation history and climatic changes in central and southern Europe. In: Harding A F (ed) Climatic change in later prehistory. University Press, Edinburgh
7. Bolle H-J (1982) Radiation and energy transport in the earth atmosphere system. In: Hutzinger O (ed) The natural environment and biogeochemical cycles. Springer, Berlin, Heidelberg, New York (The handbook of environmental chemistry 1, Part B)
8. Bradley R S, Diaz H F, Eischeid J K, Jones P D, Kelly P M, Gooders C M (1987) Science 237: 171
9. Brühl C, Crutzen P J (1988) Climate Dynamics 2: 173
10. Budyko M I (1974) Climate and life. Academic, New York
11. Budyko M I (1977) Tellus 29: 193
12. Cess R D, Goldenberg S D (1981) J. Geophys. Res. 86: 498
13. Chou M D, Peng L, Arking A (1982) J. Atmos. Sci. 39: 2651
14. Clark W C (ed.) (1982) Carbon dioxide review 1982. Clarendon, Oxford
15. Craig H (1961) Science 133: 1833
16. Crutzen P J, Brühl C, Schmailzl U (1988) Nitric acid haze formation in the lower stratosphere: a major contributing factor to the development of the antarctic "ozonehole". In: Hobbs P V (ed) Aersol and climate. Deepak, Hampton, VA
17. Cunnold D M, Prinn R G, Rasmussen R A, Simmonds P G, Alyea F N, Cardelino C A, Crawford A J, Fraser P J, Rosen R D (1986) J. Geophys. Res. 91: 10797
18. Dansgaard W (1964). Tellus 26: 436
19. Dansgaard W, Johnsen S J, Reeh N, Gundestrup N, Clausen H B, Hammer C K (1975) Nature 255: 24
20. Davies H C (1985) Climate and its variations. In: Fröhlich C (ed) Das Klima, seine Veränderungen und Störungen. Birkhäuser, Basel
21. Dickinson R E (1982) Modeling climate changes due to carbon dioxide increases. In: Clark W C (ed) Carbon dioxide review 1982. Clarendon, Oxford
22. Dickinson R E, Cicerone R J (1986) Nature 319: 109
23. Dickinson R E, Meehl G A, Washington W M (1987) Climatic Change 10: 241
24. Durst C S (1951) Climate—The synthesis of weather. In: Malone T F (ed) Compendium of meteorology, Am. Met. Soc., Boston, p 967
25. Eiden R (1979) The influence of trace substances on the atmospheric energy budget. In: Bach W, Pankrath J, Kellogg W (eds) Man's impact on climate. Elsevier, Amsterdam (Developments in Atmospheric Science 10)
26. Eiden R, Bullrich K (1987) Atmosphärische Optik: Der Massenabsorptionsindex atmosphärischer Aerosolpartikel In: Jaenicke R (ed) Atmosphärische Superenstoffe (Sonderforschungsbereiche). VCH Verlagsgesellschaft, Weinheim (For detailed information see: Fischer K (1975) J. Appl. Opt. 14: 2851 and (1976) Tellus 28: 266
27. Ewe T (1986) Das Ozon-Drama. Bild der Wissenschaft 6.86: 38
28. Faust H (1968) Das große Buch der Wetterkunde. Econ, Düsseldorf
29. Ferguson C W (1968) Science 159: 839
30. Firbas F, Losert H (1949) Planta 36: 478
31. Flach E (1981) Human bioclimatology. In: Landsberg H E (ed) General Climatology 3, Elsevier, Amsterdam
32. Flohn H (1985) A critical assessment of proxy data for climatic reconstruction. In: Tooley M J, Sheail G M (eds) The climatic scene. Allen and Unwin, London
33. Fritts H C (1971) Quaternary Research 1: 419
34. Fritts H C, Blasing T J, Hayden B P, Kutzbach J E (1971) J. Appl. Meteorol. 10: 845
35. Gates W L, Cook K H, Schlesinger M E (1981) J. Geophys. Res. 86: 6385
36. Genthon C, Barnola J M, Raynaud D, Lorius C, Jouzel J, Barkov N I, Korotkevich Y S, Kotlyakov V M (1987) Nature 329: 414
37. Georgii H-W (1985) Promet 15: 5
38. Gilliland R (1982) Clim. Change 4: 111
39. Grassl H (1986) Promet 1: 19
40. Gray J (1981) The use of stable-isotope data in climate reconstruction. In: Wigley T M L, Ingram M J, Farmer G (eds) Climate and History. Cambridge University Press, Cambridge

41. Grootes P M, Mook W G, Vogel J C, de Vries A E, Horing A, Kistemaker J (1975) Zt. Naturforschung 30A: 1
42. Hammer C U, Clausen H B, Dansgaard W (1980) Nature 288: 230
43. Hansen J, Johnson D, Lacis A, Lebedeff S, Lee P, Rind D, Russell G (1981) Science 213: 957
44. Hansen J, Lacis A, Rind D, Russell G, Stone P, Fung I, Ruedy R, Lerner J (1984) Climate sensitivity: analysis of feedback mechanism. In: Hansen J E, Takahashi T (eds) Climate processes and climate sensitivity. American Geophysical Union, Washington, DC, p 130
45. Hansen J, Russell G, Rind D, Stone P, Lacis A, Lebedeff S, Rudey R, Travis L (1983) Mon. Weather Rev. 111: 609
46. Hasselmann K (1976) Tellus 28: 473
47. Henderson-Sellers A (1986) Clim. Change 8: 25
48. Hoyt D V (1983) Variations in the solar constant caused by sunspots and faculae: an updated look. In: McCormac B M (ed) Weather and climate responses to solar variations. Colorado Associated University Press, Boulder, CO
49. Hoyt D V, Eddy J A (1982) An atlas of variations in the solar constant caused by sunspot blocking and facular emissions from 1874 to 1981. NCAR Tech. Note TN/194 + STR, Boulder, CO
50. Idso S B (1987) Climatic Change 10: 81
51. Idso S B (1987) Theor. Appl. Climatol. 38: 55
52. Jaenicke R (1980) J. Aerosol Sci. 11: 577
53. Jones P D, Wigley T M L, Wright P B (1986) Nature 322: 430
54. Jouzel J, Lorius C, Petit J R, Genthon C, Barkov N I, Kotlyakov V M, Petrov V M (1987) Nature 329: 403
55. Kellogg W W (1987) Climatic Change 10: 113
56. Kiehl J T, Dickinson R E (1987) J. Geophys. Res. 92: 2991
57. Kondratyev K Y (1972) Radiation processes in the atmosphere. World Meteorological Organization. WMO-No. 309, Genf
58. Kondratyev K Y, Moskalenko N I (1985) The role of carbon dioxide and other minor gaseous components and aerosols in the radiation budget. In: Houghton J T (ed) The global climate. Cambridge University Press, Cambridge
59. La Marche V C, Hirschboeck K K (1984) Nature 307: 121
60. Lamb H H (1977) Climate, present, past and future. Methuen, London (Climatic history and the future vol 2)
61. Lamb H H (1982) Climate, history and the modern world. Methuen, London
62. Lamb H H (1983) Climate Monitor 12: 79
63. Lamb H H, Johnson A I (1966) Secular variations of the atmospheric circulations since 1750. Geophysical Memoirs. Her Majesty's Stationery Office, London, pp 1–125
64. Land K C, Schneider S H (eds) (1987) Forecasting in the social and natural sciences. Climatic Change 11 (special issue)
65. Landsberg H E (ed in chief) (1969–1986) World survey of climatology. Elsevier, Amsterdam 15 vols
66. Landsberg H E (1981) City climate. In: Landsberg H E (ed) General Climatology 3, Elsevier, Amsterdam
67. Landscheidt T (1983) Solar oscillations, sunspot cycles and climatic change. In: McCormac B M (ed) Weather and climate responses to solar variations. Colorado Associated University Press, Boulder
68. Legrand M R, Lorius C, Barkov N I, Petrov V N (1988) Atm. Environment 22: 317
69. Leith C E (1985) Global climate research. In: Houghton J T (ed) The global climate. Cambridge University Press, Cambridge
70. Le Roy Ladurie E L (1970) Times of feast, times of famine: a history of climate since the Year 1000. Allen and Unwin, London 1971 (Doubleday, New York 1970)
71. Lough J M, Fritts H C (1987) Climatic Change 10: 219
72. Manabe S, Bryan K, Spelman M J (1979) Dyn. Atmos. Oceans 3: 393
73. Manabe S, Stouffer R J (1980) J. Geophys. Res. 85: 5529
74. Manabe S, Wetherald R T (1980) J. Atmos. Sci. 37: 99
75. Manabe S, Wetherald R T, Stouffer R J (1981) Climatic Change 3: 347

76. Manley G (1974) Q. J. Royal Met. Soc. 100: 389
77. Markus T A, Morris E N (1980) Buildings, Climate and energy. Pitman, London
78. Martinson D G, Pisias N G, Hays J D, Imbrie J, Moore T C, Shackleton N J (1987) Quat. Res. 27: 1
79. McIntyre D A (1980) Indoor Climate. Appl. Sci. London
80. Mitchell J F B (1983) Q. J. Royal Meteorol. Soc. 109: 113
81. Mitcheil J F B, Lupton G (1984) Progress in Biometeorology 3: 353
82. Mitchell J M (1963) On the world-wide pattern of secular temperature change. In: Changes of climate. UNESCO Arid Zone Research Series XX, pp 161–181
83. Mitchell J M, Dzerdzeevskii B, Flohn H, Hofmeyr W L, Lamb H H, Rao K N, Wallén C C (1966) Climatic change. World Meteorol. Organ. Techn. Note 79, Genf
84. Neftel A, Moor E, Oeschger H, Stauffer B (1985) Nature 315: 45
85. Neftel A, Oeschger H, Schwander J, Stauffer B, Zumbrunn R (1982) Nature 295: 220
86. Newman M J, Rood R T (1977) Science 198: 1035
87. Newman P A, Schoeberl M R (1986) Geophys. Res. Lett. 13: 1206
88. Nihoul J C J (ed) (1985) Coupled atmospheric-ocean models. Elsevier, Amsterdam
89. North G R, Cahalan R F, Coakley J A Jr (1981) Rev. Geophys. Space Phys. 19: 91
90. Paterson W S B, Waddington E D (1984) Rev. Geophys. Space Phys. 22: 123
91. Peng L, Chou M-D, Arking A (1987) J. Geophys. Res. 92: 5505
92. Perseke C, Schmidt H, Schirmer H (1987) VGB Kraftwerktechnik 67: 811
93. Pfister C (1985) Snow cover, snow-lines and glaciers in Central Europe since the 16th century. In: Tooley M J, Sheail G M (eds) The Climatic Scene. Allen and Unwin, London
94. Pisias N G, Martinson D G, Moore T C, Shackleton N J, Prell W, Hays J, Boden G (1984) Marine Geology 56: 119
95. Pittock A B (1986) Eos 67: 193
96. Pittock A B, Frakes L A, Jenssen D, Peterson J A, Zillman J W (eds) (1978) Climatic change and variability. Cambridge University Press, Cambridge
97. Pollard D (1982) The performance of an upper-ocean model coupled to an atmospheric GCM: preliminary results. Climate Research Institute, Oregon State Univ, Rep. No. 31
98. Potter G L, Kiehl J T, Cess R D (1987) Climatic Change 10: 87
99. Prospero J M, Charlson R J, Mohnen V, Jaenicke R, Delany A C, Moyers J, Zoller W, Rahn K (1983) Rev. Geophys. Space Phys. 21: 1607
100. Ramanathan V, Cicerone R J, Singh H B, Kiehl J T (1985) J. Geophys. Res. 90: 5547
101. Ramanathan V, Coakley J A Jr (1978) Rev. Geophys. Space Phys. 16: 465
102. Raschke E, Bandeen W R (1970) J. Appl. Meteorol. 9: 215
103. Reidat R (1981) Technical climatology. In: Landsberg H E (ed) General Climatology 3, Elsevier, Amsterdam
104. von Rudloff H (1967) Die Schwankungen und Pendlungen des Klimas in Europa seit dem Beginn der regelmäßigen Instrumentenbeobachtung (1670). Vieweg, Braunschweig
105. Schlesinger M E (1983) Int. J. Environ. Stud. 20: 103
106. Schlesinger M E (1984) Climate model simulations of $CO_2$-induced climate change. In: Saltzmann B (ed) Advances in geophysics. vol 26, Academic, New York
107. Schlesinger M E (1986) Climate Dynamics 1: 35
108. Schneider S H, Dickinson R E (1974) Rev. Geophys. Space Phys. 12: 447
109. Schönwiese C-D (1979) Klimaschwankungen. Springer, Berlin Heidelberg New York
110. Schönwiese C-D (1986) Theor. Appl. Climatol. 37: 1
111. Schönwiese C-D (1986) Meteorol. Rdsch. 39: 126
112. Schönwiese C-D (1987) Beitr. Phys. Atmosph. 60: 48
113. Schönwiese C-D, Malcher J (1987) Der anthropogene Spurengaseinfluß auf das globale Klima — statistische Abschätzung auf der Grundlage der Beobachtungsdaten. Institut für Meteorologie und Geophysik, Univ. Frankfurt (GDR)
114. Seiler W, Giehl H, Brunke E-G, Halliday E (1984) Tellus 36B: 219
115. Sellers P J, Mintz Y, Sud Y C, Dalcher A (1986) J. Atmos. Sci. 43: 505
116. Sellers W D (1969) J. Appl. Meteorol. 8: 392

117. Sellers W D (1980) Physical Climatology. University of Chicago Press, Chicago
118. Simmons A J, Bengtsson L (1985) Atmospheric general circulation models: their design and use for climate studies. In: Houghton J T (ed) The global climate. Cambridge University Press, Cambridge
119. Somerville R C J (1987) Climatic Change 11: 239
120. Spelman M J, Manabe S J (1984) J. Geophys. Res. 89: 571
121. Stephens G L (1984) Mon. Weather Rev. 112: 826
122. Stothers R B (1984) Science 224: 1191
123. Stuiver M (1978) Nature 273: 271
124. Tooley M J, Sheail G M (eds) (1985) The climatic scene. Allen and Unwin, London
125. Tricot C H, Berger A (1987) Climate Dynamics 2: 39
126. Tromp S W (1980) Biometeorology. Heyden, London
127. Untersteiner N (1985) The cryosphere. In: Houghton J T (ed) The global climate. Cambridge University Press, Cambridge
128. US GARP Committee (1975) Understanding climatic change. Nat. Acad. Sci., Washington
129. Visconti G (1985) The influence of atmospheric trace gases and aerosols on climate. In: Fröhlich C (ed) Das Klima, Seine Veränderungen und Störungen. Birkhäuser Basel
130. Wang W-C (1985) Climatological effects of atmospheric ozone: a review. In: Zerefos C, Ghazi A (eds) Atmospheric ozone. Reidel, Dordrecht
131. Wang W-C, Molnar G (1985) J. Geophys. Res. 90: 12971
132. Wang W-C, Wuebbles D J, Washington W M, Isaacs R G, Molnar G (1986) Rev. Geophys. 24: 110
133. Washington W M, Meehl G A (1983) J. Geophys. Res. 88: 6600
134. Washington W M, Meehl G A (1984) J. Geophys. Res. 89: 9475
135. Washington W M, Semtner A J, Meehl G A, Knight D J, Mayer T A (1980) J. Phys. Oceanogr. 10: 1887
136. Wetherald R T, Manabe S (1981) J. Geophys. Res. 86: 1194
137. Wetherald R T, Manabe S (1986) Climatic Change 8: 5
138. White O R (ed) (1977) The solar output and its variation. Colorado Associated University Press, Boulder, CO
139. Wigley T M L (1984) Climate Monitor 13: 133
140. Wigley T M L, Ingram M J, Farmer G (1981) Climate and history. Cambridge University Press, Cambridge
141. Williams L D (1974) Computer simulation of glacier mass balance throughout an ablation season. Proc. Western Snow Conference, Anchorage, Alaska, pp 23–28
142. Williams L D (1975) Arctic and Alpine Research 7: 169
143. Wilson R C, Gullkis S, Janssen M, Hudson H S, Chapman G A (1981) Science 211: 700
144. World Meteorological Organization (1966) Climatic change. WMO-No. 195. TP 100, Geneva
145. World Meteorological Organization (1974) Physical and dynamic climatology. WMO-No. 347, Gidrometeoizdat, Leningrad
146. Yao A Y M (1981) Agricultural climatology. In: Landsberg H E (ed) General Climatology, vol 3. Elsevier, Amsterdam

# Subject Index

# The Handbook of Environmental Chemistry

O. Hutzinger, Bayreuth, FRG (Ed.)

## Volume 2:
## Reactions and Processes
## Part D

With contributions by P. B. Barraclough,
N. O. Crossland, R. Herrmann, W. Mabey,
C. M. Menzie, T. Mill, P. B. Tinker, M. Waldichuk,
C. J. M. Wolff

1988. XI, 210 pp. 47 figs. 55 tabs.
ISBN 3-540-15547-3

**Contents:** *R. Hermann,* Bayreuth, FRG: Hydrology.
– *N. O. Crossland,* Sittingbourne, UK; *C. J. M. Wolff,*
Amsterdam, The Netherlands: Outdoor Ponds:
Their Construction, Management, and Use in
Experimental Ecotoxicology. – *T. Mill,* Menlo Park,
USA; *W. Mabey,* San Francisco, USA: Hydrolysis of
Organic Chemicals. – *M. Waldichuk,* Vancouver,
Canada: Exchange of Pollutants and Other
Substances Between the Atmosphere and the
Oceans. – *P. B. Tinker,* Swindon, UK;
*P. B. Barraclough,* Harpenden, UK: Root-Soil
Interactions. – *C. M. Menzie,* Washington, D.C.,
USA: Reaction Types in the Environment

An important purpose of **The Handbook of Environmental Chemistry** is to aid the understanding of
distribution and chemical reaction processes which
occur in the environment. Volume 2, Part D of this
series is dedicated to a broad description of chemical
reactions of pollutants in environmental compartments, to root-soil interactions and to hydrology.

Springer-Verlag Berlin
Heidelberg New York London
Paris Tokyo Hong Kong

Springer